Mermaids, Monasteries, Cherokees and Custer

the stories behind
Philadelphia
street names

Robert I. Alotta

Bonus Books, Inc.

© *Robert I. Alotta, 1990*
All rights reserved

Except for appropriate use in critical reviews or works of scholarship, the reproduction or use of this work in any form or by any electronic, mechanical or other means now known or hereafter invented, including photocopying and recording, and in any information storage and retrieval system is forbidden without the written permission of the publisher.

94 93 92 91 90   5 4 3 2

Library of Congress Catalog Card Number 90-82322

International Standard Book Number: 0-933893-90-6

Bonus Books, Inc.
160 East Illinois Street
Chicago, Illinois 60611

First Edition

Printed in the United States of America

Unless otherwise credited, photographs and illustrations are used with the permission of the Free Library of Philadelphia, Print and Photograph Collection.

For the first namer, who shall remain nameless.

Contents

Foreword—vii

Acknowledgments—xiii

Introduction—xv

Alphabetical listing of streets—1

Bibliography—251

Acknowledgments from the 1975 edition—257

Index—259

Foreword

I admit I've always been curious about "the story behind the story," about why certain things are as they are—like how streets get their names. Bob Alotta, who did the prodigious research required to produce this book, shares that kind of curiosity—and, obviously, if you find yourself drawn to its pages, so do you.

I realize there are people who could drive on Roosevelt Boulevard every day of their lives and never once wonder whether it was named for Franklin D. or Theodore—or maybe Eleanor? But I think life is more fun when you take notice of such details. (It's named for Theodore and was built long before Franklin and Eleanor moved into the White House.)

Discovering the story behind a street name is a way of learning history, of developing a better sense of a place and its people. I learned early in my career as a journalist that many people care very deeply about what a street is named—especially if they are seeking to change its name, or prevent the name from being changed.

We've had some experiences in Philadelphia in the last few years that illustrate what an emotional issue street naming can be. The best known of these is what might be called the Great Drive Debate.

For more than 100 years, the winding drives through Fairmount Park on either side of the Schuylkill River were known as the East River Drive and the West River Drive. No would be surprised to learn that the former was on the east side of the river and the latter on the west. Such names are straightforward, practical, and non-controversial.

Foreword

But then, in 1985, John B. Kelly, Jr., died suddenly from a heart attack. Kelly Jr. was a former Olympic rowing champion, a major booster of the sport of rowing on the Schuylkill, a member of City Council, a member of the Fairmount Park Commission, and a member of a particularly famous Philadelphia family—his sister was Princess Grace of Monaco, his father had once run for Mayor of Philadelphia (and, so the story goes, had the election stolen from him). Kelly Sr. also had been a rowing champ, a member of the Park Commission, and a supporter of sports on the river. A sculpture of Kelly at the oars of a scull sits on the east bank of the river.

Stunned by Kelly's unexpected demise, his colleagues on the Park Commission voted 15-0, without debate, to rename the East River Drive Kelly Drive to honor father and son together. Since the drive is totally contained within Fairmount Park, the Commission had the authority to rename it without City Council approval.

Almost immediately, an organization was formed to champion the renaming of West River Drive. The candidate for this honor was restauranteur Frank Palumbo who, until his death in 1983, had been a generous patron of the Philadelphia Zoo—which is located on the *west* side of Fairmount Park. The Park Commission received 30,000 cards and letters supporting the change. Philadelphia City Council voted unanimously in favor of it.

But, again, since this is a road within Fairmount Park, the Park Commission had the last word—and this time, the word was *no*.

The commissioners agreed that Mr. Palumbo was worthy of tribute, but by this time, they'd had a chance to hear from still another constituency—the traditionalists, who preferred the Park names to be Left Alone. The traditionalists, it turned out, weren't happy about Kelly Drive either. They weren't happy that an earlier bout of moniker meddling had eliminated the delightfully named Wissahickon Drive by decreeing it an extension of Lincoln Drive.

"Maybe in retrospect we acted too hastily in renaming Kelly drive," the Park Commission president admitted. "The Commission now has no desire to rename anything."

Foreword

And that ended name changing in the Park—for a while, at least—but it certainly didn't end citizen determination to honor by street sign. Not at all.

In the midst of the Park Drive debate, an organization in North Philadelphia petitioned City Council to officially rename Columbia Avenue Cecil B. Moore Avenue.

Moore, who died in 1979, was a city councilman, one of the city's best-known defense attorneys, and a civil rights leader who led countless demonstrations against racial discrimination. Moore's name was rarely mentioned without the word controversial or flamboyant somewhere in the sentence. And he enjoyed living up to both descriptions.

When Council appeared to dilly-dally over this renaming proposal, Moore's supporters borrowed some of Moore's tactics: they held a demonstration. And shortly thereafter, Cecil B. Moore Avenue was added to the map.

But Columbia Avenue didn't totally disappear. Council finally voted to rename Columbia only between Broad Street and 33rd—that is, in the district from which Moore had been elected to Council. It is still Columbia Avenue from the river to Broad Street, and it's Columbia Avenue for another 30 blocks past 33rd.

This solution might have dismayed the mapmakers, but it had the virtue of avoiding the Mario Lanza Problem.

You see, years ago, fans of the late Philadelphia-born singer who became famous in Hollywood wanted to have Christian Street in South Philadelphia renamed for Lanza—*all* of Christian Street. And that brought protests from sections of the street where, clearly, Mario Lanza fans were outnumbered. These Christian Streeters did not want the trouble of having their records and stationery changed. Thus Christian Street remains.

All of the above is not to say that *every* street name change involves a debate. There was general agreement, I recall, when Pennsylvania Avenue was renamed John F. Kennedy Boulevard following the President's assassination in 1963.

And, frankly, the public often doesn't even know when streets are named. Many streets are built by and named by developers—who then turn both street and name over to the city when the development is com-

Foreword

plete. In 1987, a developer of new homes in northeast Philadelphia decided to name a byway Pelle Road after Philadelphia Flyers goalie Pelle Lindbergh, who was killed in an auto accident in 1985. If City Council had attempted to name a street for Lindbergh, you can be sure that Mothers Against Drunk Driving would have picketed City Hall, since Lindbergh was intoxicated when he crashed his car. But this naming wasn't noted in the papers at the time.

Still, some people in the area noticed, and their response was interesting. They simply asked local developers to draw from a list of names they felt were more deserving—names of city police and firefighters who had died in the line of duty. In 1988, the first of these, Joseph Kelly Terrace, was dedicated.

I realize that no book could possibly contain every fact about every name on our street signs. Bob did not attempt to list all the roads within Fairmount Park. One of my favorites there is Forbidden Road—so named because automobiles are forbidden there. The name is noted on all the signs, although it's actually an unofficial one, dating back to the time when autos first became popular.

You may well find in these pages a name you know more about than Bob Alotta's succinct paragraphs convey. I felt that way when I came to Buist Avenue. "You left out the story about the poinsettias," I told Bob. "You can write about it in your introduction," he replied. So I will:

South Carolinian Joel Poinsett went to Mexico in 1825 as America's first "minister" to that country after it became independent from Spain. When he returned to the United States, he took with him a plant he thought would do well back home. He sent this plant (which he named for himself) to Robert Buist, who operated a seed company on his farm in West Philadelphia. It was Buist who cultivated Poinsett's seeds, and distributed them across the U.S. I think that's worthy of an avenue.

Anyway, even if Bob Alotta hasn't told us "everything," he tells us quite a lot. He has performed a true public service by satisfying his—and our—curiosity about the street names of Philadelphia. I've learned in these pages things I never knew before: That the Delaware River really should have been called the Hudson.

Foreword

That Callowhill Street, on which the newspaper I work for is located, is named for William Penn's second wife. (I didn't even know that Penn had remarried.) That Deal Street is named for a popular postman. And much, much more.

Thank you, Bob.

> *Rose DeWolf*
> Staff Writer of the
> *Philadelphia Daily News*
> June 1990

Acknowledgments

It's been more than 15 years since I wrote my first book about Philadelphia street names and, in that period of time, I am still as curious about how they come about as I was at first.

In this new edition, I have been able to do a favor for a friend. Jerry Post, when he reviewed the first edition of the book, said that I needed to add more names—more of the old names—to make it easier for a researcher. He suggested even adding an index. Well, Jerry, you got your wish. I've added more than a hundred old names to the list—in alphabetical order—and I've added a number of new street names: life still goes on in Philadelphia, and there are still debts to pay to constituencies, as you will see.

While I was bringing the book up to date in Harrisonburg, Virginia, I needed a great deal of help in Philadelphia. I got it from my son Peter, who acted as my eyes and ears in Philadelphia. I don't know what I'd do without him. LTC Bill MacDonald's tripod, I must say, saved the day.

An old and dear friend, and now City Councilman Thacher Longstreth let me use the good services of his legislative assistant, Jonathan D. Weinstein, who brought me up to date very quickly. And Jerry Post, who moved from the Map Collection at the Free Library of Philadelphia to head the Print and Photograph Collection, has continued to be of help. Jerry, I must note, has been acknowledged in each of my books. That must be some record. I'd also like to thank the anonymous people in Road Records at the Streets Department for their continued help. Jon Weinstein made me feel good

Acknowledgments

when he called to tell me that some of the folks there still remembered me.

It was great to research at the Free Library of Philadelphia, though it was scary to find a file labeled "Alotta, Robert I.," in Prints and Pictures, containing material I had donated to the collections. They say you can't go home again. They're wrong. You can. The only sad part is you can't stay.

And, finally, I'd like to thank all the people who made this book possible: Rose DeWolf, a dear friend and writer for the *Philadelphia Daily News,* who got the call from Owen Hurd in Chicago and contacted another old and dear friend, former *Evening Bulletin* columnist Jim Smart. He, in turn, put her in touch with my lawyer, Labron K. Shuman, who finally reached me here in Virginia.

Thanks also to Bonus Books for assigning Sharon Turner Mulvihill to act as my editor. Together we had a few laughs and delivered the book on time.

It's been great fun reliving the past . . . with a little help from my friends, even if, in all these years, they never got together and named something "Alotta Boulevard."

Robert I. Alotta, Ph.D.
Harrisonburg, Virginia
15 September 1990

Introduction

Is it to achieve some sense of belonging that many people express such interest and curiosity in knowing how the name of their street, their neighborhood, their city came into existence? Or is it an innate desire to feel that in this highly automated, depersonalized society there is some humanizing element—some token of immortality—attached to the bestowing of a name upon an inanimate object such as a street? Can this interest be the result of the revived popularity of folk history and genealogy? Or is it the wish of the people to know more about where they came from so they can better adapt themselves to where they are going? The reason may be all of these or none. But this book is the product of that very real interest.

The street names of Philadelphia have intrigued and confused folk-lorists and historians for years. A city as old as Philadelphia, so filled with tradition, has countless stories behind each and every name. And, up to a few years ago, which story to believe as the true one was a personal choice.

One could look at a name like Phil-Ellena, and decide it was named after a man named Phil, and a woman, Ellena. Or, a man named Frank who ran a ford across the Delaware River gave his name to Frankford. Or, a woman with a bustle caused the name of Bustleton to appear.

Those are just a few of the misunderstood, misinterpreted street names in Philadelphia. Phil-Ellena was the name of George W. Carpenter's mansion. He modified the Greek *philos*, love, and combined it with his wife Ellen's name. The road leading to the mansion

Introduction

finally adopted the name. Frankford was an anglicized version of Frankfurt, the company that settled the area. The family Bussil operated a tavern in what was then called Bussil-town. Again, the name was corrupted to an Americanized version.

Regardless of the accepted or desired meanings, Philadelphia has a kaleidoscope of street names, which reflect cultural affectations and the changing patterns and influences of society's thrust and parry.

There is music in the names on Philadelphia streets. Little children once jumped rope to a cadence created by a repetition of center city street names. Some even wrote poetry to celebrate the names. One such extravagant display of affection and praise follows:

> The streets of Philadelphia
> Are elegant and old.
> How sweetly sound the verdant names
> When on the tongue they're rolled.
> Of Walnut, Chestnut, Spruce and Pine,
> Spring Garden, Cherry, Locust, Vine,
> Myrtlewood and Appletree,
> Lavender and Strawberry.
>
> But hold! Like raucous cry of crow
> Sounds the name of Weccacoe.
> No chiming glocken. Tulpehocken
> Smites the ears—and Conshohocken.
>
> Horrocks! (But it is a street
> And not an exclamation)
> And Champlost—here is one defeat
> Without an explanation.
> It puts me in a dreadful funk
> To read the name of Manayunk:
> Who'd have thunk there'd be a Shunk
> Or even one named Passyunk?
>
> If you find these street names shocking
> Wait 'til you read Wingohocking,
> Wissinoming and Poquessing.
> That's not all: there's still Kingsessing.
> When thou lookest for Venango
> Doth thee note of Ogontz, Mutter.
> Take thee note of Ogle, Napfle.

Introduction

> The streets of Philadelphia
> Are elegant and old.
> How full of 18th Century grace
> The lovely name Delancey Place.
> And who is there who would not dally
> At a street named Elfreth's Alley?
>
> But when my friends ask me to tell
> Where's the house wherein I dwell,
> The house, in short, I pay a tax on.
> How I pale and look away.
> How I cringe when I must say
> That I live—to my dismay—
> On Shackamaxon.*

There is music and rhythm. But there is much more.

When an individual or group selects a name to represent some area of land, be it street or place, it is a communication of the culture in which that individual or group lives.

As with the naming of children, street and place naming goes through phases of popularity. Unlike people naming, streets and places are not restricted by family tradition. Streets and places never carry a "junior" or roman numeral as an affectation. Though unencumbered by rigid orthodoxy, street and place names are directly affected by politics, religion, and other cultural variables.

"The naming of (streets)," one Philadelphia columnist wrote, "must be a ticklish business. In new neighborhoods, we have noted a tendency to get into one category and stick there. Flowers, for instance, so that they may have a sequence of Iris, Daffodil, Primrose, Crocus, and so on. This is much safer than picking up political figures. Flowers rarely make themselves unpopular by ill-considered policies, and there is not likely ever to be a march on City Hall demanding that the very name of gladiolus be wiped from the city plan."†

It is safe to say that once emblazoned on a signpost, the name remains until the end of time—or until city

---

*Madway, Edna Bockstein. "The Streets of Philadelphia," *The Bulletin Almanac,* undated—but published sometime prior to 1960.

†Jones, Paul. "Candid Shots," *The Bulletin,* January 30, 1958.

Introduction

planners remove the street entirely, or houseclean the records of unpopular names. By the time you read this, Manderson Street might be no longer. Since it was declared officially open in 1894, Manderson Street has been systematically reduced in length to a single block. A good example of the latter was the change of Shippen Street to Bainbridge. The patriotic Commodore William Bainbridge was a mite more popular than Benedict Arnold's father-in-law, even though Edward Shippen was Philadelphia's first elected mayor.

To know why a particular stretch of land was called thus-and-such is to know something about life—and society—at a particular moment in history.

Before the first settlers arrived in this country, the earliest roadways were only rough trails blazed through the woods by the Indians. As the nation became more and more colonized, the footpaths were beaten back and widened by the increased traffic and by natural erosion. These paths were quite adequate for foot travel, but the introduction of heavily laden wagons required something more. Many times, the early colonists took it upon themselves to clear the way for a road from their home to that of a neighbor, linking it into an old Indian trail or a recently fabricated public road.

Around Philadelphia, the need for more accessible public thoroughfares was first felt by the inhabitants of areas outside the limits of Penn's "old city." They came, sometimes singly, sometimes in a group, to Quarter Sessions Court—so named because it met quarterly—and made petitions for roadways to their "plantations," or from their ferry, or to their meeting house. The petitions were heard and, if it were the court's decision that such a roadway be considered, a committee of interested citizens was appointed to view the proposed terrain. After the committee reported back to the court, its report was confirmed or denied. If the proposal was confirmed, the court issued orders for the ground to be cleared and the road constructed. The records of the surveys were then transcribed into road dockets.

As the Philadelphia area became more and more urbanized, and more and more of Penn's original grants were sold off into smaller and smaller parcels, the need arose for more sophisticated methods of regulating streets and roads. Land needed for the roadbeds was

Introduction

acquired by deed of dedication and by eminent domain. Real estate men developed an area, then dedicated the roadbed to the city. When several developers built houses along the same route, however, they would sometimes forget to deed the streets to the city. In these cases, when it became necessary for the way to be widened or otherwise altered, the city solicited the aid of residents of the neighborhood, who would swear by affidavit that the street or road had been in public use for twenty-one years or more. In the absence of long-time residents, the city solicitor rendered an opinion that effectively—and legally—said the same thing. In extreme cases, the members of city council would pass an ordinance for acquisition. Included in this ordinance was a three- to six-month clause which enabled adjacent landowners to contact the city and enter a claim for damages, if any.

After the consolidation of the districts and townships of the County of Philadelphia into the City of Philadelphia in 1854, a consolidated city plan was developed. Streets to be built in years to come were sketched onto the plan—and modified as dreams became reality. When a street, a road, or a section of either was no longer necessary, it was removed from Philadelphia's master plan; the roadway itself obliterated by housing or another road being widened, and finally abandoned for all purposes.

Streets, roads, avenues, and alleys have a character all their own—a character imparted to them by the terrain through which they travel, the ideas of the men who planned them, and the names given to them. The City of Philadelphia has never maintained a record of why particular thoroughfares were given the names they bear. Perhaps no one ever thought of this as a function of government. The earliest recorded names were most times derived from the point of destination—whether a meeting house (Orthodox Street), another town (Lancaster Avenue), or someone's farm (Buist Avenue). Nor did the namers forget the original settlements which once made up the County of Philadelphia (today contiguous with the city). Later, streets cut through the estates of prominent Philadelphians assumed the name of the estate (Phil-Ellena Street), the mansion (Belfield Avenue), or the owner (Bringhurst Street).

Introduction

There are six basic sources for street names:

Descriptive: Based on observation and experience, these names are derived from geographic features, or from characteristics drawn from the owner, residents, structures, or human or animal activities. This category contains the names William Penn wanted on all his original streets, the names of "things that grow . . . and are native" to the land. As a result, in this category, we find Chestnut, Walnut, Spruce Streets; Papermill, Bells Mill Roads; Flat Rock Road; Cliveden, Mount Pleasant, Fairmount Avenues, and others.

Descriptive names represent less than 10 percent of all Philadelphia streets.

Honorary: Names that commemorate or recognize the accomplishments of individuals who may or may not have had a direct connection to the location. In this category appear the names of Jefferson, Washington, Franklin, Wayne, Pulaski, McClellan, Meade, Bancroft, and a litany of Philadelphia's mayors and Pennsylvania's governors.

Honorary names represent almost 34 percent of all Philadelphia streets.

Nepotistic: These names demonstrate personal favoritism, which may or may not have a direct relationship to the site. Nepotistic names are usually given by housing developers who, with city permission, name the streets that course through their projects. As a result, the houses built by Joseph Bucknell front on Bucknell Street. Subsequent real estate entrepreneurs named *their* streets for their wives, children, and themselves, resulting in Lois Lane (with no apologies to Superman), Augusta Street, Ann, Larry—even Zeralda.

Nepotistic names represent little more than 10 percent of all Philadelphia streets.

Migratory: Names in this category "travel" or migrate, recalling a place or feature found in another city, state, or country. It is suggested that these names appear in an effort to eliminate homesickness. This has been documented in at least one case: Darlington Street was named for the hometown of an efficient city clerk who wanted to quit her job and return home.

Migratory names represent almost 24 percent of all Philadelphia names.

Allusions: Originating in literature, history, popular culture, etc., these names have an implied or indirect

Introduction

reference. Allusions can be found in streets such as Titan, Ivanhoe, and Nautilus.

Allusory references represent 8 percent of all Philadelphia streets.

Contrived: These names are ingenious fabrications, where the actual meaning is never obvious—except to the namer. An excellent example of this type of naming can be found in a private housing development in the Northeast section of Philadelphia, Normandy Village. The developer only assigned names that began with the letter "N," including Nestor and Norcum.

Contrivances represent almost 10 percent of all Philadelphia street names.

It should be noted that cursory category definition can provide a researcher with unreliable data which can result in undependable conclusions.

Here is an example that illustrates this point.

Greenstreet, a hypothetical name, could fall into any of the categories listed. It could be *descriptive* of the area, from the viewpoint of an early settler. It could be an *honorary* appellation, sheer *nepotism*, or the *migratory* memory of a homesick pioneer's former residence. On the other hand, it could conceivably be an *allusion* to a fictional character, or a *contrivance*: Who ever saw a street that was green?

There are pitfalls in determining street name origins, pitfalls that can be avoided by strict historical research, an overall knowledge of literature, language—all the liberal arts—and the foibles of mankind.

In the early days of America's settlement, naming a town could not be put off for an indefinite period. In some instances, such as Philadelphia, the name was decided upon before actual settlement.

William Penn's choice of Philadelphia, for "the virgin settlement" of his new land, "named before thou were born,"* described what he wanted his city to be: a "City of Brotherly Love." He could have retained the Delaware Indian name for the area: *Coaquannock*, the "Grove of Tall Pines." Or, he might have used the name of the place where the alleged Treaty Elm stood, *Shackamaxon*, but he didn't. His notion was that this new city

---

*The Penn Mutual Life Insurance Company. *Your Friend, WILLIAM PENN*, Philadelphia: privately published, 1944, p 7.

Introduction

was different, not like anything in that place before. He might also have realized that the Indian name for the treaty area was open to wide interpretation, based on different dialects of the Lenni-Lenape language. Shackamaxon can mean "Place of the Eels," or "Meeting Place of the Chiefs."

Penn's name-choice was not original. It was a migratory name. Philadelphia was named after the ancient city of Lydia whose original name, Rabbath-Ammon, was changed in the second century B.C. to honor an Egyptian of Greek descent, Ptolemy Philadelphus.*

The selection of a name for a place or a street has important ramifications, since, under normal circumstances, the name remains forever. The name adds to, or detracts from, the character of the locality. The public view and comprehension of an area is based on first impressions: how the name is initially perceived. Few people would seriously consider going to a surgeon who lived on Bloody Alley, or frequenting a fragrance shop on Dung Drive.

Names that are serious or respectable—or, at the very least, neutral—provide a better climate for future success. Penn had this in mind when he landed in his settlement and found that his well-meaning surveyor-general, Thomas Holme, had named all the streets for friends of Penn. Holme was ordered to replace those names with those of the trees found in the forest which would be Philadelphia. Holme's own name was also replaced with Mulberry, now Arch. The surveyor did get even: there are roadways now called Holmes and Holmesburg.

Because of Penn's conscious effort to neutralize envious antagonism, the original official names were descriptive. The first exception was Callowhill Street, named for Penn's second wife, Hannah Callowhill.

Prior to Penn's departure from America, descriptive names dominated. In the Old City, the downtown area, these names appeared exclusively—save for the numbered north-south streets.

In areas not under strict Quaker control, a few

---

*Burt, Struthers. *Philadelphia: Holy Experiment*. Garden City, NY: Doubleday, Doran & Co., 1945, p 43.

Introduction

streets honoring people did appear. What better way to show one's affection for a patron than by naming a street for him—or her. The feminine honorary names in that early period were sometimes disguised. A case in point: Christian Street, in what was once part of the Swedish settlement south of the City, obtained its name from Sweden's girl-queen, Christina. Queen Street, nearby, also honors her.

Other names appeared honoring the person or family that owned the plot of land on which the houses and road were built. Thus, we find the names of Fitzwater (a nineteenth-century mapmaker's spelling of Fitzwalter) Street, Elfreth's Alley, and others. But these unFriendly names were few and far between. Penn's directive was carried out until his and his family's influence waned, then finally disappeared.

The names developed prior to 1800 were limited in their association to Philadelphia, flora/fauna, and Pennsylvania. After all, there wasn't much more from which to draw.

The first half of the nineteenth century showed little physical growth in the County of Philadelphia. A Revolution had been fought; Independence won, and a Constitution ratified. It was a period of catching up. The lack of street construction during this period is really not surprising. There were no major immigrations, no vast migrations to the City/County. There were no concentrations of people which would force the building of streets.

In the second half of the nineteenth century, descriptive names began to disappear, replaced by migratory, or transplanted, names. Honorary names, however, dominated. The names of Signers of the Declaration of Independence—and Ratifiers of the Constitution—along with military officers and others appear. Darrah Street, named for Lydia Darragh, who reportedly warned Washington at Whitemarsh of British plans to attack him, appeared during this period—one of few Philadelphia streets named to honor women.

During 1851–1900, over 15 percent of all the streets which appear on modern maps were created. Growth was in the rural areas and the townships. Though they were "country," an appreciation of nature was not apparent. Less than one percent of all the new names

Introduction

assigned during that period were descriptive. The largest number honored someone or something from Philadelphia, Pennsylvania, abroad, or the United States.

Something new appeared: allusions. Based, perhaps, on the intellectual renaissance, and the advent of free, public education, names began to paint pictures or recall literary memories. And, for the first time, people began to name streets after themselves and their families. Sometimes there was good reason. Because the city government would not pave the roadway between his rowhouses—the first in Philadelphia—William Sansom named the street built at his own expense, after himself. Though very few nepotistic names appeared, it still created a precedent.

By the beginning of the twentieth century, honorary names peaked, followed by a substantial increase in migratory, nepotistic, and allusory names. Descriptive names, assigned principally in the distant areas, showed a remarkable comeback, reaching and surpassing the heights achieved during the earliest periods of Penn's settlement. As in the past period, a new type of name appeared: the contrivance. These names, having no basis whatsoever in reality, were the products of creative, idle minds.

Finally, since 1951, the last major growth period—almost half the existing streets were created after that date—the names developed into a pattern.

As time went on, Philadelphians in authority took note of the contributions made by the movers and shakers, and named streets, boulevards, and any other conceivable stretch of roadway in their honor.

Streets of minor significance, the internal streets, tend to be named in honor of a local resident, rather than a person of national importance. As a result, one can expect to find streets associated with a local postman (Deal Street), a druggist (Cantrell Street), the first baptism at a local church (Ruffner Street), or some type of foreign influence, such as a wealthy immigrant (Busti Street). This is a safe naming technique. If the street is considered insignificant, the only persons to please are those in the immediate vicinity—not members of the entire community.

Streets of major importance—those that traverse an entire city, or significant portion thereof—cannot honor

Introduction

the resident of only one portion of that street. There would be too much opposition from the people who live at the other end of the avenue. As a result, these streets carry the names of "neutral" personages, those of city-wide or national reputation, usually from the United States . . . with no foreign influences whatsoever.

The "neutral" influence can be seen in the series of Philadelphia streets with the names of presidents . . . and, the names of Pennsylvania's governors, from 1709–1895. Two governors who ignored the City of Philadelphia by withholding funding during their administrations are omitted. As one can see, naming is a two-edged sword.

Even the Indians, neglected and forgotten by countless generations, were commemorated in this way. Precious metals and stones, heroic events, authors (and some of their fictional characters), mythological symbols, states of the union, people's home towns, governors, counties, mayors, statesmen, presidents, even housepainters, were included in the naming.

During Consolidation, many historically significant street names were abandoned. But during the period following Consolidation, names that were changed were returned to their original designations. This happened in 1897–98, as a result of pressures brought to bear by civic and historical groups.

The road of research for this book was often impeded by obstacles and roadblocks. One of the principal ones, as noted before, was the lack of cohesive record-keeping by the City of Philadelphia. Some clerk, unknown and forgotten, tried to create a list with plausible reasons for the street names—but he gave up between *A* and *E*. Another obstacle, of greater proportion, was the seemingly authoritative newspaper clippings which are generously sprinkled throughout the institutions where the research for this book was conducted. The Free Library has several books of clippings from the old *Public Ledger* which, for many years, were considered gospel. In another of their files, loosely entitled "Streets," are accumulations of other clippings—these from *The Philadelphia Inquirer*, the *Evening Bulletin*, the *Record*, the *North American*, and a few anonymous newspapers. Most—if not all—of these clippings are undated and contain a potpourri of misinformation and erroneous data. In fact,

Introduction

there is no agreement or even similarity from clipping to clipping in the same newspaper. Reporters for these newspapers, in many cases, perpetuated popular misconceptions. Sometimes they did not even take the time to review the research conducted by former staffers on the same paper. In very few cases do they indicate a source.

A good example is Grant Avenue. Not one of the newspaper clippings consulted indicated that a Samuel Grant once owned the land through which this roadway passes. Each one, in turn, stated—without fear of contradiction—that the avenue was named to commemorate General (later President) U. S. Grant. The same was true with Penrose Avenue. No one took the time to find out who owned the ferry. They all assumed Penrose honored a famous Philadelphia politician—of a much later date—who had the same name. This type of data hampers research because each case requires backtracking to determine whether the popular, printed statement is accurate—and, if not, compile enough evidence to disprove the misconceptions.

A major source for this book, and one which few people consider, was the original road records, dockets, jury decisions, and surveys contained in the City of Philadelphia's Archives. Though the sheer bulk of these documents with their spidery penmanship is daunting, they provide the best possible source of information. But this information is worthless without access to the streets department's official road records. The department maintains a complete listing of every existing street in the City of Philadelphia—by block. Each entry is backed up by an official document of some sort—be it affidavit, city solicitor opinion, road docket, ordinance, or what have you—which can be found in the larger filing system in their offices.

Because of the sheer mass of more than three thousand streets, it is impossible to assume there are no errors in these records. There are. Sometimes, through clerical oversight, an ordinance is not properly listed or a survey is not included. To support a belief, one must also look into the countless volumes of city ordinances and resolutions concerning name changes and street openings and closings.

Sometimes, a clerk with foresight will attach a memorandum to a card, such as I discovered for St.

Introduction

George's Street, which filled in a blank which might never have been filled in any other way. Indeed, there were times when my research consisted of recalling the past with clerks in the streets department who, with their associations, afforded information which would not have been found otherwise.

After determining the first official date for a particular street, research becomes a drudgery of looking through countless lines of almost agate type in voluminous street directories in an attempt to locate persons with similar names who lived in the general area. Sometimes, frustration takes the upper hand, and the street is dropped from the list—not because it is unworthy of note, but because there is insufficient evidence to establish from where or whom the name was derived. With a decade and a half separating my research, I was able to persevere and add a number of names that frustrated me before. How many? I'm afraid to count.

An interesting sidelight to research on the naming of streets is that there is no official body in the City of Philadelphia which decides what names goes where. The only one who ever systematically determined names was William Penn. Present practice is to adopt any name submitted to the streets department provided it does not duplicate an existing street name and is neither controversial nor too difficult to pronounce.

One of the most useful printed works for researching street names is Scharf and Westcott's *History of Philadelphia*. The book is filled with references to the contributions made by various Philadelphians to the improvement of the city—and the dates they did what they did. Even more helpful were its descriptions of the topography and geography of Philadelphia during the early days, which assisted immeasurably in determining where the descriptive names for streets originated. Other useful books are listed in the bibliography.

The reader will notice that there are no source notes for the information given in the street entries. For the casual reader, this should make the book more readable and less intimidating. For any reader who would like to pursue a particular topic further, lists of the full documentation have been conveyed to both the Free Library of Philadelphia and the Historical Society of Pennsylvania.

By no means have all the streets of Philadelphia

Introduction

been covered in this book. Though it borders on sacrilege, it must be said that the criterion for inclusion was nothing more than the bulk of available, accurate data I was able to compile. Very early in the research phase it became necessary to decide that if nothing concrete could be found, nothing would be written.

There is something else the reader should know. The only streets listed here are those for which I feel enough is known to arrive at a logical, factual conclusion. Many others could be included which would, for the most part, be fanciful creations of gossip and hearsay.

Naming is organic. It is mankind's method of leaving memorials to society and its members, and milestones to progress. With each name, a conscious reminder of the past is created for succeeding generations.

The choice of a name is a subjective decision, based on observation, consideration of alternatives; the implications and ramifications, and public acceptance/rejection.

Why was this book written? Simply because it was there. No one had ever done it before. Street naming is an area of Philadelphia history which has been skipped over by historians and civic-minded individuals since the founding of the city. It is hoped that this book will be a contribution to the wealth of hidden knowledge about Philadelphia. It is also hoped that it will underscore the need for the city to require a statement of reason before any new street receives its name.

Abbottsford Avenue

Charles Frederick Abbot (1821–97), an affluent admirer of novelist Sir Walter Scott, came to the falls of the Schuylkill in 1845. The main reason for his visit to that then-lush area was to look for a home.

As he viewed a mansion built before the Revolution in 1752, we're led to believe, he was reminded of the novelist's home on the banks of the Tweed, back in Scotland. Immediately, he thought of the monks from nearby Melrose Abbey, hiking up their cassocks to ford the stream near Scott's Abbotsford. Tradition tells us that's where the name came from.

On the other hand, one would be surprised if Abbot had named his mansion anything else, regardless of his interest in Scott's writings.

The community which grew up around Abbotsford, now the site of the Medical College of Pennsylvania, assumed the name of the mansion. The road which led to the house also acquired the name. It adopted the beds of Wyoming Avenue and Norris or Olney Street, and added its last section in 1916. In 1941, the road became an avenue. The spelling of the name—one *t* or two—fluctuates throughout the record books. Officially, the name is now spelled Abbottsford.

Abbot served as a member of the Pennsylvania legislature from 1858 to 1862 and, from 1868 to the time of his death, as a member of the Board of Education.

Aberdeen Street

See American Street.

Abington Avenue

Abington Avenue is named for the early Quaker settlement of Abington in Montgomery County. But the road does not lead to that suburban community. A continuation of Lincoln Drive, it extends from Pastorius Park to Ardleigh Street.

The entire avenue was officially opened in 1921.

Abington Road

See Washington Lane.

Absalom Jones Way

Part of Fifty-second Street, between City Avenue and a dead end southeast of Paschall Avenue, was renamed by Philadelphia's City Council in 1983 to honor the founder (along with Richard Allen) of the Free African Society, the first black service organization in America.

In 1787, Jones led blacks from St. George's Methodist Church in protest of a segregated seating policy in the church's gallery. Seven years later, Jones founded St. Thomas Protestant Episcopal Church, the first black Episcopal congregation in Philadelphia. He later became the first black American to be ordained an Episcopal priest.

Academy Road

In the first few years of the eighteenth century, the inhabitants of the northern reaches of Penn's provinces petitioned the Supreme Executive Council to provide them with proper roadways leading to their "plantations." Their request was quickly approved by Quarter Sessions Court, and a public road from Frankford to Red Lion was surveyed and confirmed in 1710.

A tiny log school near the northern terminus of this road, built with a small legacy (£4) left in 1723 for "school purposes" by Penn's surveyor-general Thomas Holme (see Holmesburg Avenue), grew into the Lower Dublin Academy. It is from this small log cabin that the road obtained its name. In 1808, the Lower Dublin Public School was erected on the site, at present-day Academy and Willits Roads, where the Thomas Holme School now stands.

Academy Road was confirmed as a legally open street with the title of Road in 1941. Additional sections were included in the 1950s—the most recent in 1959.

Academy Street

See Appletree Street.

Acorn Street

Named, in the Penn tradition, for the starting point of a tree native to the area.

Agate Street

Another example of naming for local flora, fauna, and minerals.

Agree Alley

See Quarry Street.

Agusta Street

Opened in 1926, Agusta Street was named for Agusta Pulch, the wife of George W. Shisler, a real estate developer who laid out a portion of this street (Vankirk to Benner) and dedicated it to the city. At least two other streets, Alma (for a daughter, perhaps) and Shisler, were named by Shisler in commemoration of his family.

The mortgage on Shisler's land, for some reason, was foreclosed during the Depression. A legal opinion in 1935, however, stated that, regardless of the foreclosure, the street was legally open, and it continues to bear Mrs. Shisler's name.

The last two segments of Agusta Street were deeded to the city in 1953.

Albert Street

See League Street.

Albright Street

This street and Albright College both derive their name from Jacob Albright (1759–1808), the Montgomery County-born founder and first bishop of the Methodist movement.

Alcott Street

ALCOTT STREET. Though Louisa May Alcott, seen here with her father Amos Bronson Alcott, spent only two years of her life in Philadelphia, the city proudly remembered her as one of its own.

Named in honor of the creator of *Little Women*, Alcott Street from Saul to Penn Street came into existence in 1924.

Born at 5425 Main Street (now Germantown Avenue), Louisa May Alcott (1832–88) spent only two years of her life in Philadelphia. Then her father moved the entire family to Boston, where he taught school. Later, after they had moved to Concord, she became one of America's foremost novelists.

Alcott Street grew steadily during the 1920s. The last addition, Newtown Avenue to Weymouth Street, was deeded to the city in 1930. In 1976, the section of Alcott Street from Hawthorne to Mulberry was renamed Anchor.

Alden Street

John W. Frazier, Esq., registrar in Philadelphia's Bureau of Surveys at the turn of the twentieth century, named Alden Street in honor of one of his wife's ancestors, John Alden (of Miles Standish–Priscilla Mullen fame). Mrs. Frazier herself, the former Anna Maria Redfield, was honored by her husband's naming of Redfield Street.

Nor did Frazier overlook his own forebears in his street naming. Frazier Street took care of that.

The earliest documentation for Alden Street is in a deed presented to the city in 1903 for the roadbed from Race to Summer Street. Affidavits taken in 1916 and 1917 indicate that the street from Arch to Race had been in public use for at least twenty-one years. This would suggest that the earliest public use was approximately 1895–96. The last section to be added, from Greenway Avenue to Kingsessing Avenue, was deeded to the city in 1919.

Allegheny Avenue

Allegheny Avenue is one of several Philadelphia streets to be named after Pennsylvania counties (see Counties). A portion of the avenue appears as early as 1818 as part of old Front Street, but deeds referring to it by name document the earliest section to be from the pierhead line of the Delaware River to Aramingo Avenue. The last segment to be added was deeded to the city in 1917.

Incorporated in 1788, Allegheny County was named for the river. According to Indian experts, Allegheny is probably a corruption of *alligew-hanna*, which translates to "stream of the Alligew (or Talligew)," a tribe which, in keeping with the tradition of the Delaware Indians, once occupied the region east of the Mississippi. The definition is disputed, however, and some believe the name means "endless," "the best river," "fair water," "river of the cave," or "he is leaving us and may never return."

Allengrove Street

The main access to the Allen-Grove Female Seminary, off the main street of Frankford (now Frankford Avenue), between Wakeling and Harrison Streets, was dedicated to the city in 1894.

The street was officially opened as a "legal street" by affidavit in 1921, after the seminary had been demol-

Allengrove Street

ished. Other sections of Allengrove were added between then and 1927. In 1936, Harrison Street, from the Roosevelt Boulévard to Castor Avenue, became Allengrove. Later the same year, the last segment, from Summerdale Street to the Boulevard, was included.

The marble facade of the Frankford Historical Society and this street name are all that remain of the seminary.

Allens Lane

ALLENS LANE. The road that once led to the country seat of Chief Justice William Allen bears his name.

Sections of the road leading to the country seat of Chief Justice William Allen (1704–80) date back to 1746, as a road which "ran over to Thomas Livezey's Mill at 'Glen Fern' on the Wissahickon," though, strictly speaking, in that year the mill still belonged to Thomas Shoemaker (see Livezey Street).

Allen, a noted jurist, laid out a small settlement in the northeastern part of Pennsylvania in 1765. That town, Allentown, also bears his name.

East Allens Lane, from Sprague Street to Stenton Avenue, was known as Nippon Street prior to 1926. The growth of Allens Lane accelerated at the turn of the century, the most recent section being dedicated to the city in 1958.

Allison Street

See Appletree Street.

Alma Street

See Agusta Street.

American Street

After the 1854 consolidation of the several townships and districts which made up the County of Philadelphia into what is known today as the City of Philadelphia (see Consolidation), residents of the new metropolis were utterly confused by the duplication of street names. In an effort to create coherence out of chaos, an ordinance was passed in 1858 to unify nomenclature. Between 1858 and 1896, a determined effort was made to apply single names to continuous streets.

It was in this name-changing process that American Street came to be. From as early as 1787, this street had borne such names as Corn (Reed to Manton Street), Strongford (Christian north to a dead end), Guilford (Monroe to South Street),* Ashland and Aberdeen (Delancey to Spruce), Levant (Spruce to Chancellor), St. John (Vine to Germantown Avenue), and Washington (Cadwallader to Oxford).

The majority of blocks which now make up American Street were either deeded to the city or declared legally open by affidavit during the mid- to late nineteenth century. The last block, from 65th Avenue north to 67th, was added in 1961.

Anchor Street

The nearness of Anchor Street to Little Tacony Creek (which turns toward the Delaware River) and the old Aramingo Canal probably triggered the street namer to apply a nautical name. Dedicated to the city from 1924 to 1928, the street stretches from Colgate Street to Hegerman. In 1976, a section of Alcott Street, from Hawthorne Street to Mulberry, was renamed Anchor.

---

*Probably the oldest section. According to the 1787 street directory, a Guilford Street already existed "to near Plumb [now Monroe] street."

Anderson Street

ANDERSON STREET. Major Robert Anderson, for whom this street was named, was an unlikely hero. A Southerner by birth, and temperament, Anderson commanded Fort Sumter in Charleston Harbor when the South Carolinians fired the first shot of the Civil War.

The street commemorates Major Robert Anderson (1805–71), the unlikely first hero of the Civil War. The Kentucky-born Anderson commanded Fort Sumter in Charleston Harbor when the South Carolinians opened fire and began the Civil War. During the Black Hawk War, Anderson served with Jefferson Davis, later president of the Confederate States of America, and administered the officer's oath to a young Abraham Lincoln. When Anderson visited the City of Philadelphia upon his return from Sumter, he was feted and received the ultimate honor: a street named for him.

Andreas Street

See Camac Street.

Angora Terrace

During the Civil War, David Callahan built a woolen mill at 60th Street and Baltimore Avenue. As an inducement to his employees and their families, he also built a village on both sides of Baltimore Avenue, between 58th Street and Cobbs Creek. He called the village Angora, probably for the wool which was used in his mill.

For himself, Callahan purchased the property of David Snyder, and built a mansion at 58th Street and Baltimore at a cost of $30,000. The property included a vast expanse of woods called Sherwood Forest. The forest burned in 1900 and was leveled for homes in 1913. A year later, the mansion was demolished.

After the fire, adjacent land (from 55th Street to 57th Street) was deeded to the city. Between 1914 and 1925, the remainder—to 61st Street—was added. By a 1948 resolution of the Board of Surveyors, acting on a city ordinance, Angora Street became Angora Terrace.

Appletree Street

Appearing in the first logical street directory (1791), Appletree Street was named for the fruit trees which lined it. In 1895, to achieve uniformity of naming, several old streets became Appletree, including Allison (Front to a dead end west), Winfield (7th to 8th), Knights (or Knecht's) Court and Birch Place (8th to 9th), Academy (10th to 11th), Budden's Alley (12th to Juniper),* Kershaw (Burns to a dead end east), Grace (16th to a dead end eastward), Tower (20th to Beechwood Street), and Hall (56th to 57th).

But the original Philadelphia street called Appletree, between 4th and 5th Streets, no longer exists. Urban development removed it in 1965.

Aramingo Avenue

Gunner's Run, a stream in the old District of Northern Liberties, was the basis for the naming of Aramingo Avenue. The Indians called the stream, which has since disappeared, *Tumanaraming*, which the settlers abbreviated and corrupted to Aramingo. The original name, appearing on a 1689 deed, means "wolf walk," an appropriate designation, since the area was infested with wolves when it was first settled by the English. The area surrounding the waterway—the village of Doverville and its neighborhood—became known as the Borough of Aramingo in 1850 (see Consolidation). In 1854, the borough was incorporated into the City of Philadelphia.

There was a scheme afoot during that period to widen and deepen Gunner's Run and convert the stream into a major trade facility. On March 15, 1847, the Gunner's Run Improvement Company was incorporated, empowered to construct a canal. The work was useless. Business was insufficient to cover the expenses and the canal was considered by most to be a nuisance. Nonetheless, by act of 1850, Gunner's Run was renamed the Aramingo Canal.

---

*The section from 13th to Juniper was vacated by ordinance, January 14, 1965.

Aramingo Avenue

The earliest section of Aramingo Avenue, across Wheatsheaf Lane, dates back to 1862. Then, when the canal was stricken from the city plan in 1887, orders were give to "substitute a street of the same width." The most recent section to be added, between Sepviva and Milnor Streets, was deeded to the city in 1955. The lower end of the avenue (see Dyott Street) is said to cover the bed of the old canal.

Aramingo Street

See Tacony Street.

Arch Street

Originally designated as "Holme's Street," in honor of Penn's surveyor-general Thomas Holme, Arch was renamed Mulberry Street by Penn, for the numerous mulberry trees which dotted the area. The last of these trees were removed in 1917.

The earliest survey of the lots of land on Mulberry was made in 1683, from 3rd Street to 4th, on the north side. The last of the provincial surveys took place in 1740 on the north side, between 7th and 8th (see Original Streets).

In April 1690, Benjamin Chambers, Thomas Peart, and Francis Rawle presented a petition to the Provincial Council "that a bridg might be built over and a wharfe made against Mulberry St." In order to open Mulberry to the Delaware River, it was necessary to cut through a hill or knoll in the vicinity of Front Street. This left one end of the street higher than the other. An arch or bridge was needed to carry Front Street over Mulberry.

Built before the end of 1690, the "arch" was sixty-six feet long and was considered a great engineering feat. The people of Philadelphia referred to it as "the great arch" and the street, "the arch street." The arch was torn down in 1721, but the name continued in common use for more than a century. In 1853, the name was officially changed to Arch Street.

By affidavit rather than deed, Arch Street, from the Delaware to the Schuylkill, was incorporated as a "legally open" street in 1883.

ARCH STREET. Originally called Holme's Street—then Mulberry—this street was the site of one of America's first engineering wonders: "the great arch." The arch, or bridge, was created to compensate for a slight discrepancy in street building: when Mulberry was opened to the Delaware River, they had to cut through a hill near Front Street. This left one side of the street higher than the other; *ergo*, the arch! (Illustration is a mythic creation to resemble what the arch might have looked like in the 1700s. Note William Penn and an Indian at the lower right-hand side of the artwork.)

Ark Lane

See Stone House Lane.

Armat Street

Opening through the meadow of the Armat family estate in Germantown, the once rural lane assumed the name of the family. More specifically, it received its name from the Armat Mills, which stood on the street.

Armat Street

The Armat family came to this country about the time of the Revolution and settled in Loudon County, Virginia. After the war, they emigrated to Philadelphia. Then, along with countless other Philadelphians, they moved from the city proper to escape the yellow fever epidemic of 1793.

Germantown was to their liking, and in 1801, Thomas Armat (1749–1831) began construction of a home for his son, Thomas Wright Armat, adopting for it the name of the Virginia county where he first lived. Loudon Mansion, still standing, lent its name to Loudon Street.

The young Armat was a distinguished merchant and philanthropist. In his spare time, he faced the "energy crisis" of his day—the rapidly decreasing supply of firewood. He was one of the first Americans to suggest the use of coal as a source of heating fuel.

In the mid-nineteenth century, the City of Philadelphia attempted to change the name of the street to Maplewood. In 1897, the name returned to Armat.

Armstrong Street

Although folk historians have contended, erroneously, that this street was named for the Pennsylvania county of the same name, Armstrong derives its name from a nineteenth-century mill in Germantown. The street appears on early maps as "a lane running to [Danenhower's] mill . . . called . . . the 'Road to Shellebarger's Mill.'" An 1871 atlas indicates that John Armstrong, whose mill was situated near Wister's Woods, owned almost all the land between Shoemaker Station and Duy's Station along the Wingohocking Creek, where the street now runs.

Armstrong Street's earliest section, formerly known as Mercer, from Rubicam to a dead end, was stricken from the city plan in 1972. The oldest remaining section is from Belfield Avenue to Wister Street.

Ash Street

See Fletcher Street.

Ashburner Street

Charles Albert Ashburner (1854–89), Philadelphia geologist, and one of the organizers, in 1873, of the Engineers Club of Philadelphia, was the probable namesake of this street.

Ashland Street

See American Street, Bouvier Street.

Ashmead Street

John Ashmead (1648–88) and his family came to this country in 1682. He and Tobias Leech, his brother-in-law (see Fox Chase Road), purchased a fairly large tract of land in what became Cheltenham Township. The road or lane leading through the property naturally assumed the name of the landowner. The county of Montgomery carries the name of Ashmead's family home in England.

John Ashmead died accidentally in 1688. The shock of his death, it is said, precipitated his wife's death the following day. His survivors permitted Count Zinzindorf to open the first Moravian school in the colonies at Ashmead House in the spring of 1742; the school was moved to Bethlehem, Pennsylvania, in June of the same year. One of Ashmead's descendents, William Ashmead, a blacksmith, is credited—along with the Bringhurst family (see Bringhurst Street)—with creating a lightweight carriage to replace the heavier imports.

Ashmead Street first appears in an official city street directory in 1857 as "S, from Clinton, Gtn." Though it is listed, it was not an "official" street until 1866, when a section of road from Germantown Avenue to Wakefield Street was dedicated to the city. At that time, many Ashmeads—prominent lawyers and doctors, soldiers and merchants—were still living on or near the family's property. The last official segment of the street, from Magnolia to Clarkson, was deeded to the city in 1924.

Asylum Road

See Adams Avenue.

Asylum Street

See Delancey Street.

Atlantic Street

See Bancroft Street.

Audubon Avenue

John James Audubon (1785–1851), noted ornithologist and naturalist, spent a portion of his life in the Delaware Valley because "the wild woods about Philadelphia offered so many opportunities for tramping and nature investigation." The avenue which bears his name is mainly a product of the mid-twentieth century.

John James Audubon roamed the "wild woods about Philadelphia" because they provided so many opportunities for catching sight of flora and fauna.

Autumn Street

See Bouvier Street.

Avondale Road

See Leiper Street.

Axe Factory Road

The unique name of this road, one of the earliest in the Northeast, is derived from an axe factory which opened on the side of the road in the early nineteenth century. Virtually isolated, the inhabitants of the area petitioned for a road "up said Creek [Pennypack] to works now erecting by William Maghee, thus crossing said Creek to Bustleton and Springfield Turnpike." The road jury confirmed what they called Factory Road—from Bustleton to Welsh Road—in December 1811. The axe factory later became the site of a yarn mill, but the road remained unchanged until the 1950s, when it was widened.

Bach Place

A section of Manning Street that ran from 15th Street to Broad, next to the Academy of Music, was renamed Bach Place in 1985. As one streets department official said, "It seemed fitting." It might have seemed more fitting to name it after one of the Philadelphia Orchestra's legendary conductors, like Eugene Ormandy.

Bache Street

See Carlisle Street.

Bailey Street

A strong-willed woman, who wouldn't allow her sex or the death of her husband to stand in the way of providing for her small children, was honored by the city she served by the naming of Bailey Street.

In 1808, Lydia R. Bailey (1779–1869) found herself a widow, deeply in debt, and with four small children to feed. Her husband Robert had been a printer, the son of a noted Philadelphia and Lancaster printer. After his death, Mrs. Bailey carried on the business. With the help of her in-laws and her own skills as a printer, her business prospered. From 1830 to 1850, she was the city printer for Philadelphia—the first and only woman to hold such a position. In her obituary, Mrs. Bailey is spoken of as "one who enjoyed a woman's right to the full, though living before a formal exposition of that doctrine, and who as a practical printer had considerable deserved local fame." She is buried in the family vault at Old Pine Street Presbyterian Church at 4th and Pine Streets.

Bailey Street existed, "unofficially," during the Civil War in the block from Jefferson Street to Columbia Avenue. Later the street grew to include the blocks from Columbia to Montgomery Avenue. Engleside, to the south, from Parrish to Poplar, had its name changed to Bailey in 1895. From that time to 1925, the street grew to its present distance.

Bainbridge Street

The lane through the lands of the first elected mayor* of the City of Philadelphia, Edward Shippen, was laid out as a public highway by the Commissioners of Survey under the Act of February 29, 1787. The Supreme Executive Council confirmed Shippen's Lane three years later. The original section, from Passyunk Avenue to 4th Street, was added to in the early nineteenth century, and the street now runs from river to river.

Shippen's Lane became a street at the time of Consolidation. But the name "Shippen" was not held in very high esteem by most Philadelphians and many Americans. Shippen's daughter, the vivacious and cunning Peggy, was married to Benedict Arnold, and she joined her husband in the disgrace of his treason. The Shippen family is not completely forgotten in Pennsylvania lore, however. The town of Shippensburg commemorates Edward.

It was not until 1870 that Philadelphia legislated to change the name and give the street some modicum of respect. The new name, Bainbridge, commemorated a native son, William Bainbridge (1774–1833), who rose through the ranks of the early American navy to become a commodore. At Tripoli, Bainbridge captained the frigate *Philadelphia*. A few years later, during the War of 1812, he commanded a squadron of famous American fighting vessels, including the U.S.S. *Constitution* (Old Ironsides), the *Essex*, and the *Hornet*. Bainbridge is buried at Christ Church.

Records indicate that only one section of Bainbridge Street has ever been altered. That section, between 3rd and 5th Streets, was widened in 1834.

BAINBRIDGE STREET. Commodore William Bainbridge was a bonafide American hero who proved himself in battle during hostilities with the Tripolitan pirates, and the War of 1812. (Circa 1835 painting from one in the National Portrait Gallery)

Baldwin Street

Matthias William Baldwin (1795–1866), for whom this street was named, produced Old Ironsides, the "parent" of American locomotives, at his works here in Philadel-

---

*Shippen was the first mayor of Philadelphia not to be appointed by the provincial governor. He was elected by his fellow council members.

phia in 1832. Used on the Philadelphia, Germantown & Norristown Railroad, Old Ironsides was the first steam engine to run in Philadelphia. Unfortunately, the locomotive could not operate in damp or rainy weather. In inclement weather, the iron horse had to be hauled by its flesh-and-blood counterparts.

The Baldwin Locomotive Works was located on the west side of Broad Street, south of Spring Garden, on the present site of the State Office Building. The works were removed in 1921, and a statue of Baldwin, once stationed as a silent witness to the hustle and bustle of the works, was moved as a "temporary" measure to the west side of City Hall—where it stands today.

The earliest section of Baldwin Street, from Wilde Street to Silverwood, was declared legally open in 1884. The remainder was deeded to the city in 1923 and 1924.

BALDWIN STREET. A birds-eye view of Baldwin's Locomotive Works gives us an idea of the magnitude of Matthias W. Baldwin's empire.

Baltimore Avenue

Depending on the source, Baltimore Avenue started either in 1811 or in 1872—or earlier. The sixty-plus year gap is due to the fact that the original stage route to Baltimore and Washington, laid out by the road jury in 1811 and built by the Philadelphia, Brandywine and New London Turnpike Company "over the road leading from the Schuylkill to Darby, commonly called the Woodlands Road [now Woodland Avenue], where said road diverges from the Philadelphia and Lancaster Turnpike," no longer exists as such.

It is probable that the road existed earlier than 1811 as the Darby Road from Merion Meeting to Darby Meeting, built as soon as the two places of devotion were completed. But official City of Philadelphia records do not reflect these early roadways in the cataloguing of the sections of Baltimore Avenue. The first confirmation of the present-day avenue is for the Philadelphia, Brandywine and New London Turnpike Road, dated 1872, from 39th Street to 52nd Street. An affidavit on file indicates, however, that the roadway was in existence from 52nd to 61st Street prior to that 1872 date.

Baltz Street

Baltz Street, only a single block in length, was named for the J. & P. Baltz Brewing Company, which operated on 31st Street near Thompson during the Civil War. Before entering the brewery business, the Baltz brothers operated a tavern at 304 N. 3rd Street as early as 1851.

Though Baltz Street was dedicated to the city in 1888, an affidavit indicates that the street was probably in use as early as 1884. There was a time when Philadelphia starred as a "Brewery-town." W.C. Fields, in one of his noted quotes, referred to the city as "a great town for breweries."

Bancroft Street

The secretary of the navy who planned and established the United States Naval Academy at Annapolis was honored by the city which lost the first academy.

George Bancroft (1800–91), for whom Bancroft Street was named, served as secretary of the navy from 1845 to 1846. During his short term, he removed the embryonic naval school for officers from Philadelphia and re-established it in Annapolis, Maryland.

Despite this affront to its civic pride, the City of Philadelphia named a street to honor him. Appearing in directories as early as 1874, from Wharton to Dickinson, it did not begin to elongate until 1895. The extension was created by changing the names of Atlantic, south from

BANCROFT STREET. The first naval academy was located in Philadelphia, but George Bancroft, as secretary of the navy (1845–46), moved it to Annapolis, Maryland. Nonetheless, Philadelphia honored him with a street name.

Wharton; Eleanor, from Reed to Wharton; Lindsay, from Fitzwater to Bainbridge; and Millbourne, from Clearfield to Lippincott. The additional blocks which cause Bancroft Street to extend from Oregon Avenue, despite several interruptions, to Wingohocking Street were incorporated into it before 1910.

Bancroft is remembered not for his "theft" of Philadelphia's naval academy, but for his *History of the United States*.

Bank Street

This little street, not much more than a one-block alley, caused heated controversy in the late nineteenth century, when various civic groups protested against changing the name of Bank Street because it was such a historic name. Unfortunately, no one has ever been quite sure why the name is so historic.

Bank Street, where Benjamin Franklin began his street cleaning reforms, was basically a residential street —no financial institutions and just two inns, the Boar's Head and the Blue Ball. Originally called Elbow Lane, from Market Street south with a right angle turn west to 3rd Street, it then became Whitehorse Alley. When finally opened through to Chestnut Street, the whole, including Whitehorse Alley, was named Bank Street. And it is as Bank Street that it was listed in the street directory of 1804.

In 1895, Bank Street was changed to American Street. Two years later, because of "petitions and protests against the change of names . . . by the Historical Society of Pennsylvania, the Civic Club of Philadelphia and other parties," the name was returned to Bank.

The most probable reason for the name was the proximity of the street to Dock Creek to the south. Perhaps at one time, the bounds of this street were in fact the banks of the creek.

Baring Street

Named for the first Baron Ashburton, Alexander Baring (1774–1848), this street emerged in official documents two years after his death. The earliest section of the street, from 39th to 42nd Street, was deeded to the city in 1850. Seven years later, the street extended from 31st Street to 42nd—almost its present length.

Alexander Baring represented the American interests of his family's London mercantile house, Baring Brothers. While in Philadelphia, he married Anne Louisa, a daughter of William Bingham. The Baring estate, purchased from the Binghams in 1804, was called Lansdowne (see Lansdowne Avenue). In 1866, Baring's executors conveyed the estate and the ruins of the building to the city for "public use only." Active in diplomatic circles, Baring allied himself with Daniel Webster to settle the Maine-Canada boundary dispute. The result was the Ashburton-Webster Treaty of 1842.

Barkley Street

See Delancy Street.

Barnes Street

The name of Barnes Street commemorates a Philadelphian who attended the deathbeds of two American presidents. Joseph K. Barnes (1817–83) was born in Philadelphia, studied medicine under Dr. Thomas Harris of the navy, and received his medical degree from the University of Pennsylvania in 1838. Two years later, he was appointed to the medical department of the army. During the Mexican War, Barnes served with General Winfield Scott in every engagement until the capture of Mexico City.

During the Civil War, Secretary of War Edwin M. Stanton took a liking to the doctor and made him acting surgeon-general in September 1863. A year later, the "acting" was removed. As surgeon-general, Barnes attended both President Lincoln and Secretary of State William H. Seward after the shooting of the two men in April 1865. He also was one of the surgeons who fought to save the life of President Garfield during the president's last hours.

Barnes Street

The section of Barnes Street from Hartel Avenue to Borbeck Avenue was deeded to the city in 1922. The remainder, from Righter Street to Henry Avenue, was deeded in the 1940s and 1950s.

Barr Street

See Lycoming Street.

Barton Street

The most likely source for this name is Benjamin Smith Barton (1766–1815), physician and naturalist, and the nephew of David Rittenhouse (see Rittenhouse Square). There is, however, a possibility that the street honors Clara Barton, founder of the American Red Cross.

BARTON STREET. Clara Barton, founder of the American Red Cross, might have been one of the few women for whom Philadelphia named a street. Her accomplishments, however, far exceeded those of many men whose names appear on street signs today.

Bartram Avenue

The self-taught American botanist John Bartram (1699–1777), whom Linnaeus called "the greatest natural botanist in the world," lent his name to this thoroughfare.

In 1728, Bartram purchased a tract of wilderness in Kingsessing (now part of West Philadelphia) at a sheriff sale, and proceeded to transform it into a "garden of delight." The house and garden, located at 54th Street and Elmwood Avenue, is still standing and open to visitors. An attempt is made to continue Bartram's idea of displaying examples of a variety of trees, vegetables, flowers, and other plants.

Though Bartram contributed greatly to botanical science in the eighteenth century, it was not until the turn of the twentieth century that the street was named in his honor. The section of Bartram from Essington Avenue to Island Avenue was dedicated in 1970.

Bass Street

See "Things that Grow . . ."

Bayard Street

The probable origin of this street name is the family of James Ashton Bayard (1767–1815), composed of statesmen, lawyers, and diplomats.

Beach Street

In 1808, the Quarter Sessions Court confirmed a roadway leading from the Cohocksink Creek (see Canal Street) to Berks Street. A year later, the road was extended to Willow. By 1824, Beach Street had reached another stream, Gunner's Run (see Aramingo Avenue and Dyott Street). The street was completed to its full distance by the addition of the section from Schirra Drive to Cumberland Street in 1967.

Though official city streets records do not reflect it, there were earlier names for Beach, including Oak Street and Washington Avenue, from Willow Street north to the "High Bridge" (Kensington), and Hall Street, northeast of the Cohocksink Creek and Oak Street. Washington Avenue became Beach Street in 1858. Beach Street, however, is listed as such in the 1804 street directory.

As its name suggests, the roadway led to a beach on the Cohocksink.

Beaumont Street

See Cobbs Creek Parkway.

Beaver Street

See Counties and Governors.

Beck Street

See Delaware Avenue.

Beckwith Street

See Sartain Street.

Belfield Avenue

Although the mansion of the "beautiful field" dated back to 1708, the road leading to it did not become public until 1875.

Belfield Mansion, once occupied by the famous artist Charles Willson Peale, came into the Wister family in 1826, when Sarah Logan received "Bellefield" as a wedding present upon her marriage to William Wister. The mansion stayed in that family's possession until the house was demolished.

In 1875, the portion of the present-day Belfield Avenue from Penn Street to Church was called Peale. The section from Haines to Chew was known as Cedar Lane before 1890. In 1895, Bellfield or Bellefield Street came into being. By 1916, the entire street from Logan to Chew and from Johnson to Sprague was dedicated to public use. By resolution of the Board of Surveyors in 1913, the spelling of the street's name was unified to Belfield.

Bellevue Street

Presumably named after the town of Bellevue on the outskirts of Pittsburgh, Bellevue Street existed from Francis Street to Wylie earlier than the Civil War. This original section no longer exists. In 1898, however, Wensley Street from 20th to 22nd Street became known as Bellevue. Wensley, though not listed in the official directories, had been declared open "for at least 21 years" by 1878.

The most logical suggestions for the naming were the scenic view and the fact that a portion of the street, since vacated, led to the tracks of the Pennsylvania Railroad's right-of-way to Pittsburgh and Chicago.

Belmont Avenue

The country seat of a Tory father and a patriot son gives us Belmont Avenue.

Belmont Mansion, the "beautiful hill," is still standing in Fairmount Park and commands the view that caused William Peters to name it such in 1742. Peters, judge of the Court of Common Pleas, returned to England at the outbreak of the Revolution, but his son Richard served as commissioner of war. In 1786, the son became owner of the mansion and lived there until his death in 1828.

Richard Peters was one of the founders of the Philadelphia Society for Promoting Agriculture and "from his farm at Belmont came many model things." The property, however, was always in disrepair. When asked about this, he would comment, "How can you expect me to attend to all these things when my time is so taken up in telling others how to farm?"

The Belmont property became part of Fairmount Park in 1867. Three years later, the road jury (see Toll Roads) confirmed Belmont Avenue, from Lancaster Avenue to Westminster. By deed of dedication the same year, the avenue was extended to its present terminus, City Avenue.

The estate also gave name to the 1853 district of Belmont in West Philadelphia. The district scarcely got into business, however, before it was incorporated into the City of Philadelphia under the Consolidation Act of 1854 (see Consolidation).

Ben-Gurion Place

David Ben-Gurion was Israel's first prime minister after it became independent in 1948, and a worldwide leader in the Zionist movement. A part of Commerce Street, between 4th and 5th Streets, was renamed in his honor in 1987, a year after the centennial of his birth.

Ben-Gurion, born in Plonsk, Russian-Poland, immigrated to Palestine in 1906, where he lived out his life. In 1915, he traveled to the United States and helped form the Hechalutz to train prospective immigrants to Palestine. He was a key organizer of the General Organ-

BEN-GURION PLACE. David Ben-Gurion, one of modern-day Israel's founding fathers, was one of the most recent honorees of Philadelphia's City Council. Sadly, the Jewish leader who helped finance the American Revolution, Haym Salomon, never was so honored.

ization of Jewish Labor in Palestine, a group instrumental in forming the Nation of Israel. David Ben-Gurion has been called the Father of Israel.

Benezet Street

Anthony Benezet (1713–84) was a Philadelphia philanthropist and author who, in 1765, helped the FrenchHuguenots who escaped religious oppression in Canada. Several Acadians, as they became known, are buried in the graveyard of Holy Trinity Church on Spruce Street. One Acadian is remembered in Henry Wadsworth Longfellow's memorable poem, "Evangeline." A corruption of Acadian which refers to members of the group who migrated to Louisiana is "cajun."

Benjamin Franklin Parkway

One of the more obvious street origins, this was named for Philadelphia's "man for all seasons," Benjamin Franklin (1706–90).

Bensell's [Bensil] Lane

See School House Lane.

Berks Street

Named for the Pennsylvania county of the same name (see Counties), Berks Street, from Belgrade to Wildey, was confirmed by the road jury in 1807 as Vienna Street. By 1825, Vienna Street extended to Beach Street. More than sixty years later, it grew to Frankford Avenue.

Chatham Street, listed in the directory of 1854, ran from Broad Street to Front Street, on the same line as old Vienna. Chatham became Berks in 1858, a little more than a hundred years after the county had been named after Berkshire, England, where the Penn family held a great deal of land.

In 1901, Vienna Street was changed to Berks. Between the changing of Chatham and Vienna, Berks Street had grown to the distance from 33rd Street to Beach. From 1910 to 1921, the street was completed to its present configuration. During the middle of the twentieth century, the segment from 11th Street to 16th was stricken from the city plan and vacated.

Bethlehem Pike

In April of 1793, a petition was presented to the legislature for the establishment of a turnpike from Chestnut Hill, through Germantown, to Philadelphia. After reviewing the suggestion, the committee added that, "if carried through to Bethlehem," such a road would "be beneficial." It was recommended that the road should be built to Chestnut Hill first.

The plan was temporarily abandoned, but the Chestnut Hill and Springhouse Turnpike was in existence by the mid-nineteenth century, from Germantown to Chestnut Hill. The turnpike (see Toll Roads) was freed from toll in 1904, and the entire roadway from Lynnebrook Lane to Stenton Avenue and from Germantown Avenue to Newton Street became known as the Bethlehem Pike in 1931.

Betton's Lane

See Manheim Street.

Bigler Street

Named for a Pennsylvania governor (see Governors), William Bigler (1814–80), who was active in the promotion of the 1876 Centennial, Bigler Street attained its present length by the time of the Sesquicentennial.

Governor Bigler started out in the newspaper business with the *Clearfield Democrat*, "an eight by ten Jackson paper intended to counteract the influence of the seven by nine Whig [sic] paper." By 1841, the young Bigler was elected to the state senate, where he served six years, twice as speaker. He was elected governor in 1851. Renominated in 1854, he was not re-elected due to the activities of the Know-Nothings. Two years later, he was elected to the U.S. Senate. Bigler was a prominent figure in the Pennsylvania constitutional convention of 1872–73 and in the Centennial. His last years were spent as a railroad promoter in Clearfield, Pennsylvania.

Bigler Street appeared as an official street, from 13th to 20th Street, before World War I. The remaining blocks were added from 1920 to 1926.

BINGHAM STREET. William Bingham was a Philadelphia-born banker and legislator. He was also the first president of the Philadelphia & Lancaster Turnpike Corporation. (Scharf and Wescott's *History of Philadelphia, 1609–1884*)

Bingham Street

William Bingham (1752–1804) was a prominent Philadelphia merchant and real estate speculator, a close associate of Robert Morris, and the first president of the Philadelphia and Lancaster Turnpike Corporation (see Lancaster Avenue).

Married to Anne Willing, Thomas Willing's daughter, Bingham was a strong Federalist, and his style of living reflected this philosophy. The Binghams' home, built near Third and Spruce Streets, was modeled after the London home of the Duke of Manchester. The house and its formal gardens occupied most of the land north to Willing's Alley and west to Fourth Street.

Ann Bingham commissioned Gilbert Stuart to paint the full-length painting of George Washington that she later presented to Lord Lansdowne. That painting has become known as the "Lansdowne portrait."

Birch Place

See Appletree Street.

Birch Street

Thomas Birch (1779–1851) was one of the most prolific painters of his day. Many of his illustrations of pioneer landscape and marine life are all that remain of some elements of Philadelphia's historic past.

Black Horse Alley

This small alley, running from Front to 2nd Street, takes its name from the signboard of the old Black Horse Tavern, which stood on the alley during the first half of the eighteenth century. Though opened by affidavit in 1883, the alley apparently had been in public use for more than a century before. In fact, the Black Horse Tavern advertised live animal displays as late as 1805 at that location. Prior to being called Black Horse, the alley was called Yower's or Ewer's. In 1895, the city changed the name to Ludlow, but the original name of Black Horse was restored in 1897.

BLACK HORSE ALLEY. The Black Horse Inn or Tavern was a popular watering spot in early 18th century Philadelphia. As late as 1805, the establishment advertised "live animal displays." Illustration shows the yard of the inn/tavern, 352-354 North 2nd St.

Black Lake Road

See McNulty Road.

Blair Street

Presumably named after Blair County, Pennsylvania (see Counties), Blair Street was probably named shortly after the formation of the county in 1846. The county is the only one in Pennsylvania to commemorate a local resident—John Blair of Blair's Gap.

According to streets department records, the section of the street from Norris to Trenton Avenue was "dedicated by agreement with the district of Kensington." The district became a part of the consolidated City of Philadelphia in 1854 (see Consolidation).

The roadway from Oxford Street to Montgomery Avenue, then known as Leib Street, was in use during the Civil War. Warder and Garden Streets, from Montgomery to Norris, were in use about the same time. All three streets became incorporated into Blair in 1895.

Blessed John Neumann Place

See Saint John Neumann Place.

Blessed John Neumann Way

See Saint John Neumann Place.

Blodgett Street

See DeKalb Street.

Bloody Alley

See Noble Street.

Bockius Lane

See Haines Street.

Bonaffon Terrace

See Buist Avenue.

Bonaparte Court

A small section of Manning Street, running from 9th Street to Hutchinson, was renamed Bonaparte Court in 1985. On the northwest corner of old Manning and 9th street sits the house that once was home to Joseph Bonaparte after he fled his throne as King of Mexico.

Bond Street

See Camac Street.

Bonsall Street

Benjamin Bonsall (1687–1752), the sixth son of Richard Bonsall, inherited 104 acres of his father's land in the old neighborhood of Kingsessing. The street honors the family name.

The earliest record of Bonsall—by that name—was in 1895. At that time the street stretched from Wharton Street to Oakford. By the end of that year, however, Bonsall had grown to include several other streets, such as Hazelwood (Carpenter Street to a dead end), Cope (Locust to Chancellor and Walnut to Ionic), Mission

Bonsall Street

(Cherry to a dead end, vacated in 1919), St. Davids (Race to Vine), and Garside (Ridge Avenue to 23rd Street). By 1921, Bonsall Street had achieved its full distance.

Boone Street

Daniel Boone (1734–1820), the famous Indian fighter, was born near present-day Reading, Pennsylvania. A legend in his own time, Boone was part of General Edward Braddock's unsuccessful attempt to capture Fort Duquesne from the French in 1755; created three settlements in Kentucky, one of which became Boonesboro; helped blaze the Wilderness Road over Cumberland Gap through the Alleghenies; was kidnapped and adopted by the Shawnee Indians; served in the Virginia legislature; and was sheriff and deputy surveyor of Fayette County, Virginia.

Botanic Avenue

Once leading into Bartram Gardens, the first botanic gardens in the United States (see Bartram Avenue), Botanic Avenue has lost most of its nearness to that showplace of the past—but not its name. The earliest sections of the avenue, now vacated—from 74th Street to Island Avenue—were deeded to the city in 1890. The newest length of Avenue, from 90th Street to 92nd, was dedicated in 1932.

Boudinot Street

Elias Boudinot (1740–1821), as president of the Continental Congress, signed the alliance with France and the peace treaty with England. Under the federal Constitution, he served as a member of Congress from 1789–95. He succeeded David Rittenhouse as director of the U. S. Mint in 1795, and served for ten years. His sister married Richard Stockton (Stockton Street), who was Benjamin Rush's father-in-law. Boudinot married Stockton's sister Hannah.

The street that presently bears his name existed before 1874, at least from Kensington Avenue to Clearfield Street. But there was also another Boudinot, now 39th Street, which ran from Market Street to Elm in West Philadelphia. Originally called William, after one of the sons of Andrew Hamilton, it became Boudinot in 1858. Though the name had been changed to honor Boudinot, the 1864 street directory listed it as "or 39th Street."

Present-day Boudinot Street, running from Kensington Avenue to the Roosevelt Boulevard, was completed to its full length by 1930.

Bouvier Street

During the years of the Kennedy administration and "Camelot," folk historians tried to create some link between the first family and the City of Philadelphia. The best they could do was to tie Mrs. Jacqueline Kennedy's ancestors with Bouvier Street—after a nineteenth-century cabinetmaker, Michael (Michel) Bouvier.

Unfortunately, the tie was wrong.

Bouvier Street, from Oregon Avenue on the south to Elston Street on the north, was named for a prominent French Quaker lawyer and author, John Bouvier (1787–1851). The destitute Bouvier family immigrated to Philadelphia in 1802, where fifteen-year-old John obtained employment as a printer. Studying law at night, he was admitted to the bar in 1824. Later, he served as city recorder and associate judge of the Court of Criminal Sessions. He authored several definitive works on law, including *Institutes of American Law* and a *Dictionary of Law*.

The original Bouvier Street, from Master Street to Montgomery Avenue, was deeded to the city in 1860. Other streets along the same line had their names changed to Bouvier in 1895. These included Lingo (Roseberry to Ritner Street), Milden Hall and Lingo (Ritner to Jackson), Lingo (Passyunk Avenue to McKean Street), Lambdin or Lingo (Tasker to Reed), Lentz or Lingo (Reed to Wharton), Lingo (Washington Avenue to Carpenter Street), Ashland or Fillmore (Pine to Delancey), Illinois (Rittenhouse Square to Locust Street), Autumn (Winter to Vine), and Florence (North Street to Fairmount Avenue). The earliest segment of the present Bouvier Street is that formerly called Ashland or Fillmore, which was dedicated to the city in 1850.

Bowman's Lane

See Queen Lane.

Braddock Street

This street bears the name of General Edward Braddock (1695–1755). Braddock commanded British forces in North America in the early days of the French and Indian War. He led the abortive mission to capture Fort Duquesne (present-day Pittsburgh) from the French. He was mortally wounded in the French and Indian ambush and died. Members of his mission included Daniel Boone and a young colonial officer, George Washington.

Bradford Alley

This alley in the "old city" existed as early as 1805. Its name honored Andrew Bradford, the first Philadelphia newspaper publisher (1719). Later called Cullen, its name was changed to Naudain in 1895. The final change took place in 1971, when that block of Naudain became Bradford Alley again.

Bradford Street

Bradford Street, first dedicated to the city in the 1920s, from Napfle Avenue to Strahle Street, was named after Bradford County, Pennsylvania (see Counties), which commemorated William Bradford, second U.S. attorney general. The remainder of the street, from Roosevelt Boulevard to Napfle Street and from Strahle to Woodward, was dedicated after World War II. The sections from the Boulevard to Harbison, from Knorr to St. Vincent, from Cottman to Bleigh, and from Strahle to Lexington were vacated at the time the new segments were added.

Brandywine Street

Though folk historians have alleged that this street was named after the Brandywine Creek, there is a distinct possibility that its name commemorates the 1777 battle of Brandywine.

Brandywine Street first appeared as a street under this name before 1835, from 13th Street to Broad. The road jury viewed the street "from 13th street to 100 feet east of Broad" the next year. But it was not until 1852 that the street was confirmed. Regardless of the legality of its status, Brandywine Street did exist in public use.

Not until 1858 did Brandywine begin to grow. In that year, it picked up Centre, from 20th to 22nd Street. At nearly the same time, it acquired the sections from 32nd to 34th, from 31st to 32nd, and from Broad to 20th. In 1895, Brandywine expanded to include Rockland, from 34th to 41st; Gold, from 22nd to Judson; Wistar, from 10th to 12th; Depot, from 8th to 9th; and Minerva, from 7th to Franklin. All these predecessor streets dated back to the mid-nineteenth century.

Brewster Avenue

Reputedly named for Benjamin Brewster (1816–88), Brewster Avenue, from Island Avenue to 74th Street, was deeded to the city at the turn of the twentieth century. That entire section was relocated to its present site in the 1970s. The remainder of the avenue, from 92nd Street to Island Avenue, is the product of early twentieth-century planning.

Brewster, a Philadelphia lawyer, served as Pennsylvania's attorney general from 1867 to 1868 and as U.S. attorney general, under President Chester A. Arthur, from 1881 to 1885.

Bridge Street

Since this street was one of the first approaches across the Frankford Creek, it was logical to call it Bridge.

In 1811, Joseph Kirkbride was authorized by an act of the Pennsylvania Assembly to erect a bridge over the creek where "his ferry is now kept." Kirkbride either built the bridge or had it built. Near his ferry house, a village, sometimes called Point-No-Point, grew. Later, the community came to be known as Bridesburg in his honor. He also gave his name to Kirkbride Street.

By 1808, a roadway was viewed by Quarter Sessions Court "from Bustleton turnpike to Tacony road." Later, when the bridge was completed, the road was opened to the bridge. Bridge Street was extended in 1823 and completed to its present distance by 1930.

Brill Street

The J. G. Brill Company, "originators and builders of the first system of independent electric motor tracks for surface cars," operated at a plant on Frankford Avenue during the late nineteenth century. The company's contribution to the economy of Frankford and to American transportation is noted in the naming of this street.

The earliest segments of Brill Street date back to the Civil War, as Franklin and Taylor Streets. The street did not emerge under the name Brill until 1895. The remainder of the street was deeded to the city in the 1920s.

Bringhurst Street

At the southeast corner of what is now Bringhurst Street and Germantown Avenue stood the mansion of the Bringhurst family. John Bringhurst, along with William Ashmead (see Ashmead Street), was the inventor of the lightweight "Germantown Wagon." Because of their efforts, Germantown was an important carriage-building center during the nineteenth century. The firm's most famous customer was George Washington. Bringhurst was also a founder of Germantown Academy (see School House Lane).

Opened by affidavit, Bringhurst Street was in public use before 1864. By the turn of the twentieth century, the street extended from Laurens Street, assuming old Comlyn Street, to Rufe. In 1904, the section from Magnolia to Belfield was deeded, only to be stricken from the city plan a little more than sixty years later. The last addition to Bringhurst was in 1924, from Clarkson to Magnolia.

Broad Street

The "wide (or broad) avenue midway between the rivers" was one of the major arteries of Philadelphia as determined by William Penn (see Original Streets). Penn's initial interest in Philadelphia streets was limited to the city, so Broad Street existed in the early days as a road from Cedar (now South Street) to Vine.

BROAD AND ARCH STREETS. A 1914 view of a Clean-up Week Parade. There was a time when the City fought hard to live down the image of "Filthy-delphia." (Philadelphia Municipal Archives)

After the Revolution, Broad Street grew. First to the north: The road jury confirmed the extension of the street to Ridge Road in 1811, and six years later, from Ridge to Callowhill. Then, to the south: In 1819, the road was confirmed from South Street to Dickinson. By the middle of the nineteenth century, Broad extended along its present track from Government Avenue to Butler Street. Its northernmost extensions were created between 1903 and 1923.

For the most part, Philadelphians will state without reservation that Broad Street's twelve-mile length makes it "the longest straight street in the world." Unfortunately, even if the street didn't jog around City Hall, the longest straight street in the world would still be Chicago's Western Avenue—a 23½-mile straightaway.

Bryan Street

George Bryan (1731–91), a lawyer and politician who defeated Benjamin Franklin for a seat in the state assembly in 1764, is the most likely source for this street name.

Bryn Mawr Avenue

Built as an approach to the 1876 Centennial grounds with "12' 6" strips on each side for ornamental planting and embellishments," Bryn Mawr Avenue was named for the Montgomery County town to which it ultimately leads. The Philadelphia segments of the avenue were in use by 1876. Improvements were made before the beginning of the twentieth century.

The name of Bryn Mawr, Pennsylvania, was only a few years old when the avenue was constructed. Joseph Lesley, secretary of the Pennsylvania Railroad, had dubbed the new railroad station at Humphreysville "Bryn Mawr" in 1869, after the estate of Rowland Ellis, an early Welsh settler in the area. The Humphreysvillers changed the name of their village to tie in with the station in 1861. The name is Welsh for "high hill." Down the road apiece, the residents of Athensville, noting the new name approved by their neighbors, changed their town's name to Ardmore—"high hill" in Irish.

Bucknell Street

Bucknell Street commemorates a local resident and philanthropist and the university which bears his name. William Bucknell (1811–90) had very little formal education, but through judicious management of his money—and the profits of real estate speculation—he went on to become a multimillionaire. The University of Lewisburg (now Bucknell) was the recipient of much of his generosity. The school's name was changed in 1887 as a "thank you" for his gifts.

Bucknell Street, from Aspen to Brown, was the first section to carry his name, before 1864. This was probably one of his first areas of real estate development. The street grew north to Berks Street before 1892. In 1895, old Devon, from Wharton to Oakford, and Owen, from Buttonwood to Nectarine, were assumed into Bucknell. The remainder, from Ritner to McKean and from Oakford to Federal, was added by 1921.

Budden's Alley

See Appletree Street.

Buist Avenue

Robert Buist (1802–80), manager of the Edinburgh Gardens, came to Philadelphia in 1828 and founded a seed company, which carried his name. His farm, Bonaffon (which lent its name to Bonaffon Terrace), was located in the section of Philadelphia through which Buist Avenue now runs.

The earliest recording of Buist Avenue is an 1887 deed for the roadbed from Ashford to 83rd Street. The avenue was complete to its present distance—61st Street to 84th—by 1925.

Burbank Road

Luther Burbank (1849–1926), the noted horticulturist, is remembered in this Philadelphia roadway. After reading Charles Darwin's *Variation of Animals and Plants Under Domestication*, Burbank began a lifelong career of hybridizing plants, to produce new varieties, or to combine the best elements of several varieties in one plant. His most famous development was the Shasta daisy.

Burholme Avenue

Burholme Avenue is named for the park through which a section of it runs—from Cottman Avenue to the line of Mayfield Street. Burholme Park, the Ryerss family estate, adapted its name from the Waln family estate, Burnham, in England. The house, built in 1859 and still standing in the park, was given to the city in 1905 by Robert Waln Ryerss. The family also gave its name to Ryers Avenue (with one s).

Although the section of road through Burholme Park was vacated in 1955, it continues to be in use. The earliest section on record—from Hartel to Oxford Avenue—was dedicated to the city in 1893. The remainder was acquired between 1917 and 1923.

BURNSIDE STREET. Major-General Ambrose E. Burnside was responsible for the Union debacle at Fredericksburg, Virginia, and the death of countless young soldiers. Despite his military errors, he became famous for the side whiskers he sported: the "side burns."

Burnside Street

Ambrose Everett Burnside (1824–81), a Civil War general, was a victim of circumstances—and his own lack of judgment—during the War. At Fredericksburg, Virginia, he ordered thousands of Union soldiers to their deaths at the stone wall. He was the subject of a military court of inquiry, ordered by General George Gordon Meade (Meade Street), regarding his actions at the siege of Petersburg, Virginia. The court found Burnside at fault and he subsequently resigned his commission. He was later elected governor of Rhode Island and represented that state in the U. S. Senate until his death. However, Burnside is probably best remembered for the style of side-whiskers he wore—now known as sideburns.

Busti Street

Busti Street came into existence in 1913. It commemorates Paul Busti (1749–1824), an Italian merchant who erected a mansion on Market Street at about 44th Street in Mantua Village. Busti's property was purchased from his estate by the Pennsylvania Hospital. It became part of the hospital's Institute, originally called the Department for the Insane.

Busti is one of the few Philadelphia streets to bear an Italian name. (See also Mantua Avenue.)

Bustleton Avenue

See Toll Roads.

Butler Street

See Counties.

Byberry Road

See Toll Roads.

Cadwallader Street

This street is probably named for the Cadwalader family, whose noted members include John (1742–86), general in the Revolution; John (1805–79), lawyer; Lambert (1743–1823), colonel during the Revolution; and Thomas (1707/08–99), physician. These members used only one *l* in their names.

CADWALLADER STREET. John Cadwalader was a Revolutionary War officer who, despite Washington's urging, retained rank as a brigadier general of the Pennsylvania militia rather than accept a Continental Army commission. He and his brother, Dr. Thomas Cadwalader, donated their name to Cadwallader Street. Notice that the family members only use one *l*. (Scharf and Westcott's *History of Philadelphia, 1609–1884*)

Callowhill Street

William Penn espoused the concept of naming all Philadelphia streets after things which were native to the area. The first departure from that order was Callowhill Street, which was named for Penn's second wife, the former Hannah Callowhill (1664–1733). It is to Penn's credit that his own designation for that street in 1690 was New Street.

The road jury confirmed this street as Callowhill in 1770, extending from the Delaware River to Old York Road. It was increased in distance in 1786 and again in 1834. By 1924, Callowhill Street extended to its present length, from Delaware Avenue to Lansdowne Avenue. One part of the street, primarily that now occupied by a high-rise apartment complex, Park Towne Place, was vacated as part of Callowhill in 1959.

Penn himself is remembered in several Philadelphia street names, including Penn Street, Pennsgrove Street, and Penn Square, which, though called a square, is a street running from 15th to Juniper. Pennsylvania Avenue, the major part of which is now called Kennedy Boulevard, adopted its name not from the commonwealth but from the railroad of the same name.

CALLOWHILL STREET. The first street to depart from Penn's mandate to name streets after things that grew and were native to the land, Callowhill Street was named for Penn's second wife, Hannah Callowhill Penn. (Original painting at Blackwell Hall, County Durham, England)

Camac Street

Captain Turner Camocks (1751–1830) arrived in Philadelphia in 1804 to supervise his wife's estates. Mrs. Camocks, the former Sarah Masters, was the sister of Mrs. Richard Penn. On his trip to this country, the Captain's name underwent a sea change to Camac. The family lived at Woodvale, in the vicinity of the present 11th and Berks Streets. The Woodvale property was commonly called Camac's Woods. The grandson of Mrs. Camac was the 1859 founder of the Philadelphia Zoological Society and its first president.

Earlier predecessors of Camac Street, notably Leiper (Clover to Ludlow) and Jacoby (Cherry to Summer) Streets, date back to the early nineteenth century. The earliest segment to carry the name Camac was located between Oxford Street and Montgomery Avenue prior to 1861. This street was vacated by the city in the 1960s.

Over the years, other streets have been absorbed into Camac. These include Pallas (Shunk to Morris), formerly Bond (McKean to Moore); Dean (Lombard to Walnut);* Andress (Mt. Vernon to Melon); Duane (Brown to Parrish); Dean (Dauphin to York), originally known as Iseminger; and Cherry (Atlantic to Erie). Though not listed in the official city records, other streets which were absorbed by Camac include Mul-

CAMAC STREET. "The Little Street of Clubs," as Camac Street was once known, is the only palindrome street in Philadelphia. (Artwork by Frank Taylor, 1913)

---

*Though the official records of the Philadelphia streets department do not indicate a "Hazel alley" preceding Dean's alley, between Lombard and Walnut, the 1813 street directory does list such a street. If the directory is accurate, this would be the earliest location of Camac Street.

vaney (north from Master) and Mulberry (north from Jefferson). The street's last segment—still existing—from Bigler Street to Oregon Avenue, was deeded to the city in 1923. Camac Street is unusual, because it is the only Philadelphia street whose name reads the same way backwards and forwards.

Cambria Street

Named after the Pennsylvania county (see Counties) incorporated in 1804, Cambria Street first appears in official city streets records in 1850, from Frankford to Trenton Avenues. Subsequent additions were included from 1868 to 1911. In 1906, another street, William, from Cedar Street to the Delaware River, had its name changed to Cambria.

The name for the street, the county, and the township is derived from Cambria Hills in Wales, where many of the early settlers originated. Cambria is an ancient name for Wales. Etymologically akin to Cumberland, it means compatriots or comrades, a name the early Welsh attributed to themselves—and still do.

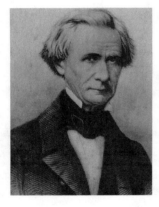

CAMERON STREET. The "Czar of Pennsylvania Politics," Simon Cameron served as Lincoln's first secretary of war. His 19th century influence in the state's political life continued into the 20th century.

Cameron Street

Simon Cameron (1799–1889), the "Czar of Pennsylvania Politics," was, among other things, Lincoln's first secretary of war. He also served as minister to Russia for Lincoln and was elected to four terms in the U.S. Senate. During his lifetime, Cameron was a newspaperman, a constructor of canals, a merger of railroads, a banker, and a very rich man—but always a political power. According to Alexander K. McClure, an old political foe, "There is not an important complete chapter of political history in the State that can be written with the omission of his defeats or triumphs, and even after his death until the present time [1905] no important chapter of political history can be fully written without recognizing his successors and assigns in politics as leading or controlling factors." He was also described in these terms: "His methods were often circuitous, the means employed were often questionable, but the end in view was always clear."

Cameron Street existed during the politician's lifetime. In 1895, the name of the street from Francis to Vineyard was changed to Dubree. Apparently, Cameron's "successors and assigns in politics" made their feelings known, because Dubree was renamed Cameron two years later.

Campbell Street

See Fulton Street.

Canal Street

The Cohocksink Creek was laid out as a public thoroughfare, with the name Canal Street, in 1795, and confirmed by the Quarter Sessions Court two years later. However, almost one hundred years went by before the legal street was an actual one.

In order to convert the creek into a street, it had to be laid with culverts and then covered. As late as 1862, the creek appears on official city maps, with a bridge over it. In 1863, a city ordinance was passed to construct another bridge, on the line of Delaware Avenue, then Beach Street.

Though official city records do not reflect it, it is probable that Canal Street became an actuality in the late nineteenth century. The street, from Laurel to Allen Street, was widened by deed in 1929.

Cantrell Street

Cantrell Street first appears in the official street directories in 1870. As it exists today, the street is mainly the product of mid- and late-nineteenth-century planning. The last segments, from 22nd to 23rd and from 28th to 29th, were included in the 1900s.

A small family of professional men, living at 1000 South 2nd Street, apparently provided such an influence on the surrounding community that this street was named in their honor. The first Cantrell (also spelled Cantrel) was John H., a druggist who conducted his business at 2nd and Carpenter Streets as early as 1847. Later, the pharmacist was joined by a Dr. John H., probably a son. By 1870, the doctor had moved a block north, to 943 South 2nd, and was replaced at the old address by Francis J. Cantrell, a lawyer.

Capewell Street

See Fletcher Street.

Carbon Street

See Counties.

Carey Street

Mathew Carey (1760–1839), though born in Ireland, was a Philadelphia publisher and author. In 1779, after writing a pamphlet protesting the British government's treatment of Irish-Catholics, Carey escaped to Paris where he met another colonial printer, Benjamin Franklin. He returned to Ireland and again got into hot water with the authorities. In 1784, he sailed for America, landing in Philadelphia with letters from Franklin and a loan from the Marquis de Lafayette. With Lafayette's money, Carey began publishing the *Pennsylvania Herald,* the *Columbia Magazine,* and the *American Magazine.* His journals were unprofitable, so he turned to book publishing. He reprinted several European books and published American authors, including Parson Mason Locke Weem's *Life of George Washington.* Active in civic life, Carey's most important contribution was the writing of articles for the Philadelphia Society for the Promotion of National Industry, establishing himself as a proponent of the nationalist school of economics.

Carlisle Street

Carlisle Street shares its name with the county seat of Cumberland County, which was named by Nicholas Scull, the provincial surveyor-general, in 1751, for Carlisle, the county seat of Cumberland, England.

Though the street name Carlisle was an official designation before 1850, segments of the street date back much earlier, but under other names, including Wetherill (Lombard to Pine), Kelton (Race to Cherry), and Bache (Vine to Race). Sections of Carlisle, from Vine to Cherry, were vacated between 1964 and 1970. Tiernan Street, from Tasker to a dead end south of Ellsworth, became a part of Carlisle in 1895—at which time Wetherill, Kelton, and Bache were joined into one continuously named thoroughfare.

Carpenter Lane

In the early days of this city, Carpenter Lane was nearly the dividing line between Germantown and Cresheim. The lane dates back to 1761, when it extended from Cresheim Road to beyond the Wissahickon Creek. Then known as Carpenter Street, the entire length of roadway, from Germantown Avenue to Wissahickon, was legally opened by affidavit in 1885. The stretch from Sherman Street to Wissahickon Avenue was widened in 1902. In 1915, it officially became a lane, to differentiate it from Carpenter Street in South Philadelphia.

The donor of the name for the lane has been said to be George W. Carpenter, a prominent druggist and scientist of the mid-nineteenth century. By the time Carpenter became well known in Philadelphia, however, the street was already an old one.

The most logical choice is Samuel Carpenter, an English Quaker and one of the first purchasers of land from William Penn. His tract of land on Holme's map of 1682 closely approximates the location of the lane. But though the lane was named for Samuel, George wasn't slighted. His family is remembered by a street named for his country manor (see Phil-Ellena Street).

Carpenter Street

See Carpenter Lane.

Castor Avenue

The original Castor Road was cut through the farm of John George Castor (or Gerster), a native of Basel, Switzerland, who came to Philadelphia in 1736. Ten years later, he moved to Germantown. In 1762, he purchased 202 acres in Oxford Township.

Though Castor's road was viewed by the road jury in 1785, it was not until half a century later that the roadway was confirmed, from Oxford Road to Asylum Road (now Adams Avenue). From 1926 to 1929, Castor Avenue grew to its present extent, adding Wyoming Street (from Unity to Adams), N Street (from Erie Avenue to Cayuga Street), and old Erie Avenue (from Delaware Avenue to present-day Erie).

John George Castor's descendents were notable Philadelphians. They included Jacob, an aide to Lafayette during the Revolution; Thomas, the inventor of the double-decker horse carriage; Elwood, also a carriage-maker; and Horace W., a prominent Philadelphia architect of the early twentieth century. Other notable members of the family, to whom past historians have given credit for the name, were General George Castor, who purchased Tacony Farm, near Frankford Arsenal, and George Albert Castor, a U.S. congressman during the early twentieth century.

Catharine Street

Opened before 1787, the original Catharine Street, extending from the Delaware River to Passyunk Road, was confirmed, at least on paper, by the Court of Quarter Sessions in 1790. In 1807, however, a traveler through the Southwark area noted that "none of the streets below South street running westward [are] laid out beyond 5th street, and Catharine and Queen streets . . . only . . . as far as 2nd street." Nine years later, the road jury confirmed the street from Passyunk to the present-day 15th Street. By mid-nineteenth century, it is safe to assume that Catharine did, in fact, extend that distance —and a little further, to Grays Ferry Avenue. When Kansas Street (Schuylkill Avenue to Bambrey Street) was changed to Catharine in 1895, the street achieved its present length, from the Delaware River to the Cobbs Creek Parkway.

Elusive as most first-name streets are to determine, it is safe to assume that the Catharine of this street was the daughter of Swen Shute (later Swanson), for whom Swanson Street was named, who settled the Swedish Southwark area on an early grant from Queen Christina (see Queen and Christian Streets).

Cathedral Avenue

See Gravers Lane.

Cayuga Street

See Indian Tribes.

Cecil B. Moore Avenue

Cecil B. Moore was a flamboyant Philadelphia lawyer who died in 1979. He was always nattily dressed, with a flaming cigar in hand and a glass of Jack Daniel's not too far away. The black civil rights activist had led the successful drive to integrate Girard College, and represented poor, black clients when no one else could—or would. Longtime president of the Philadelphia NAACP, Moore ran for mayor on the Political Freedom Rights ticket in 1967. He received 9,018 votes out of 721,348 cast. Later, Moore was elected to City Council to represent central North Philadelphia and much of Center City.

Support for naming a street in Moore's honor surfaced after the East River Drive was renamed Kelly Drive. Radio talk-show host Georgie Woods, "the man with the goods," gathered 10,000 signatures and threatened to hold a mass demonstration if City Council did not change the name.

Six months later, Council acted, and a section of Columbia Avenue, from Broad Street to 33rd, was renamed in Moore's honor. That was not what the Cecil B. Moore Ad Hoc Memorial Committee wanted, but it was all it got. Ironically, Cecil B. Moore Avenue only appears in Moore's old district. From the Delaware River to Broad and from 33rd west, it's still Columbia Avenue.

Cedar Lane

See Belfield Avenue.

Cedar Street

See Darrah Street, South Street.

Cemetery Avenue

Cemetery Avenue derives its name from its location next to Mount Moriah Cemetery, which contains the grave of Betsy Ross, interred under the name Elizabeth Claypole.

Originally called Stump, the avenue was in use by 1868, from Woodland to Chester Avenue. Surprisingly enough, there is no record in the street directories for either Stump or Cemetery in this location during that period.

Centennial Street

See Gratz Street.

Centre Street

See Brandywine Street, Counties, Rittenhouse Street, Willow Street.

CENTRE SQUARE. One of Penn's original "green spaces," Centre Square was at the center of his city. Originally used as a military drill field, the square later became the site of the city's municipal water works. In 1825, Centre Square became "Penn Square" in honor of the city's founder. By 1868, Penn Square had been cleared away to make room for Philadelphia's City Hall. During Philadelphia's renaissance in the mid-20th century, the Pennsylvania Railroad and others built skyscrapers near the square and called the development "Penn Center."

Champlost Avenue

While visiting the chateau of the Count de Champlost in 1780, George Fox was taken ill. His French host ministered to him and nursed him back to health. The Count's kindness caused the Philadelphian, upon his return to this country, to name his estate in the Count's honor. The manor house and the estate are gone, but the avenue continues to commemorate that act of kindness.

Stretching from B Street to 21st, Champlost Avenue came into being in 1888–89 as a five-block strip from 9th to Broad Street. The remainder came about after the turn of the century. The last segment added was in 1959, in the block from 20th to 21st Street.

Chatham Street

See Berks Street.

Chaucer Street

This street, which runs from Hartranft to Hulseman,

Chaucer Street

was named for Geoffrey Chaucer because, as the streets department people noted, the developer wanted a cultured, distinguished literary name for the street in his development.

Cherokee Street

See Indian Tribes.

Cherry Alley

See Cherry Street.

Cherry Street

In the early nineteenth century, many prominent Philadelphians, including Dr. James Meade and Stephen Girard, raised cherry trees from the Delaware to Front Street, along what they called Cherry Alley. The earliest record of jury confirmation of the alley is 1809, from 10th to 11th Street. The present extent of street line was deeded in the mid- to late nineteenth century. The alley became an official thoroughfare in 1883, when Philadelphia legislators decided to legitimize all then-existing streets. Apparently, many of the original public ways and private streets were never officially recorded. Cherry finally became a street in 1895, when other streets along its line had their names changed. These included Cherry Alley (Delaware Avenue to Front Street), Fetter's Lane (Bread Street to 3rd), and Howell (32nd to 34th). But one section of Cherry Street, the block from Front to 2nd, has reverted to its former name—Elfreths Alley (see Elfreths Alley).

Chester Avenue

The Philadelphia portion of the road which leads to the Delaware County town of Chester is Chester Avenue.

The present avenue is a product, on the main, of late-nineteenth-century planning.

The name of the town to which the avenue leads through Philadelphia was originally Upland. The English settlers, many of whom had come from Chester, England, petitioned William Penn to change the name. Penn, born in the same town, complied.

Chestnut Street

Originally named Wynne, in honor of Thomas Wynne, Penn's physician (see Wynnefield Avenue and Wynnewood Road), the street was changed to Chesnut (the early spelling) upon Penn's second visit to the colony (see Original Streets).

In the District of West Philadelphia, the continuation of this street was known as James Street, for James Hamilton, owner of Hamilton Village, the first settlement west of the Schuylkill. The section retains Hamilton's name, as does Hamilton Street. After the 1854 consolidation (see Consolidation), James became Chestnut.

The "old city" section of Chestnut Street had been in use from the time of Penn. The "suburban" segments are of more modern creation: 30th Street to Woodland Avenue, 1828; 41st to 44th Street, from the west line of Hamilton Village to the west line of the District of West Philadelphia, 1853; 42nd to 56th Street, 1873; 56th to 60th Street, 1897; and 61st Street to Cobbs Creek Parkway, 1906–11.

Chew Avenue

The dividing line between the town lots and the back lots of Germantown was originally referred to as Division Street. That street later became Chew Avenue.

Named for Benjamin Chew, an eighteenth-century chief justice of the Pennsylvania Supreme Court and attorney general for the state, Chew Avenue first emerged in official records as Thorp's Lane, from Wister and Penn Streets to Old York Road, in 1796. Benjamin Chew established his country seat in Germantown

CHEW AVENUE. Benjamin Chew was chief justice of the Pennsylvania Supreme Court in the 18th century. He established his country seat in Germantown and called it "Cliveden." Chew Avenue, originally called Division Street, was the dividing line between the town lots and the back lots of Germantown. (Scharf and Wescott's *History of Philadelphia, 1609–1884*)

53

Chew Avenue

about 1761 and called it Cliveden (see Cliveden Street). He and his family owned most of the land along the present route of the avenue.

The earliest section to be called Chew was recorded in the plan of the Borough of Germantown in 1849. This extended "from its intersection with Penn street, to the Bristol Township Line road." After consolidation in 1854, Chew Street grew rapidly. In 1856, the street ran from Penn Street to Washington Lane. By affidavit, the sections from Penn to Sedgwick Street were opened officially about the same time. The section from Sedgwick to Mount Airy Avenue, and from Mascher to 5th, was added before the turn of the century.

By 1931, Chew was an avenue and achieved its full length, by deed, ordinance, and the inclusion of Thorps Lane, from Ogontz to Wister Street.

CHRISTIAN STREET. Another example of chauvinism in Philadelphia street-naming is this South Philadelphia street that bears a man's name but honors a woman: Queen Christina of Sweden. (The illustration is from a portrait of the queen drawn when she was a young girl, about the time of the settlement of "New Sweden.")

Chippewa Road

See Indian Tribes.

Christian Street

Commemorating a beautiful and gracious woman, Christian Street conjures up the days of the Swedish colonization of Philadelphia before the arrival of William Penn. Peter Minuit named a natural stone wharf he passed near the mouth of the Brandywine Creek, and the fortress he built there in 1638, Fort Christina, honoring the girl queen in Stockholm.

Christian Street was opened previous to 1787 from Delaware Avenue to Passyunk. It was then known as Hudson's Lane. It is probable the name became Christian at the time of the jury confirmation. By 1837, the street extended as far west as Gray's Ferry. After the beginning of the twentieth century, Christian Street was completed to its final terminus—Cobbs Creek Parkway.

Church Alley

See Church Street.

Church Street

The street on which Christ Church stands is aptly named—Church Street.

Christ Church was founded November 15, 1695. The original building was enlarged first in 1711 and again before 1718. It is likely the street was in use at that time. Though it was listed in 1791 as a street in public use, not until 1912 was it declared officially open.

Originally named Church Alley, it became a street in 1864. From 2nd to 3rd, it was known as Jones Alley until 1866. In 1895, the entire street was renamed Commerce. Two years later, pressure by historical groups restored the original name. (See also: Phil-Ellena Street, Water Street.)

CHURCH STREET. The street on which Christ Church was built is called, of all things, "Church Street"... one of Philadelphia's unusual names. (From *Birch Views*)

Civic Center Boulevard

With extensive development to meet the increasing growth of convention business, the City of Philadelphia modernized and improved its old Convention Hall into an attractive, full-service civic center. To set off the new and renewed structures, the name of the approach was also changed. Convention Avenue, from 34th Street and Curie Avenue to South Street, and Curie Avenue, from University Avenue to the intersection of 34th Street and Convention, became Civic Center Boulevard in 1966.

The old Curie Avenue, formerly called Vintage, was dedicated on January 22, 1940, with a ribbon cutting by Mlle. Eve Curie, the daughter of the famed Marie Curie.

Clarion Street

See Counties.

Clark Street

See League Street.

Clearfield Street

Clearfield Street was named for the Pennsylvania county (see Counties) formed in 1804. According to the Rev. John Ettwein's journal, the county was named for "Clearfield Creek, where the buffaloes formerly cleared large tracts of undergrowth, so as to give them the appearance of cleared fields. Hence, the Indians called the creek Clearfield."

Clearfield Street came into existence in 1867, from 2nd Street to Rosehill, and was complete by the turn of the twentieth century.

Cleveland Street

(Stephen) Grover Cleveland (1837–1908), for whom this street was named, was the twenty-second and twenty-fourth President of the United States. He was—and is—the only president to serve two non-consecutive terms of office.

Clinton Street

"Clinton street," a 1936 *Public Ledger* editorial praised, "is the last street of downtown Philadelphia to retain, in appearance as in spirit, the atmosphere and sterling qualities of yesterday."

Named for DeWitt Clinton, builder of the Erie Canal, the block from 9th to 10th Street was originally the property of the Pennsylvania Hospital; the block from 10th to 11th belonged to the Philadelphia Almshouse. Though city records do not reflect the first actual date of the street's use, it is safe to assume it was traveled over daily at least by 1850.

In 1895, Clinton Street's name was changed to Delancey. Two years later, the name was returned to Clinton.

Cliveden Street

The Chew family mansion, which served as a focal point in the 1777 Battle of Germantown, gave its name to Cliveden Street. The house, built between 1763 and 1764 and still standing, was named for the English estate of Frederick Lewis, Prince of Wales.

Cliveden Street, long in use as an access road to the mansion and estate, was formally entered as a legally opened street in 1893. Two years later, the street included old Norton (Park Line Drive to Emlen Street). In 1923, Upsal Street (Chew to Stenton) was changed to Cliveden. The remaining blocks from Stenton to Limekiln Turnpike were deeded by 1940.

Clymer Street

The most likely origin of this street name is George Clymer (1739–1813), a Philadelphia delegate to the Continental Congress who signed the Declaration of Independence.

CLYMER STREET. George Clymer was a signer of the Declaration of Independence and a Philadelphian. (Philadelphia Municipal Archives)

Coates Street

See Fairmount Avenue.

Cobbs Creek Parkway

Cobbs Creek Parkway carries a very old name for a fairly modern roadway.

Adjacent to Cobbs Creek Park, the parkway was built in the early twentieth century after the City Council authorized the establishment of the park in 1904. Complete to its present distance and width by 1926, the Parkway assumed old Gray's Lane (confirmed in 1750) to the north side of Baltimore Avenue near 59th Street; Beaumont Street (name changed in 1924), from 65th Street north to the parkway; and Federal Street, from 61st Street to Washington Avenue.

Cobbs Creek, whose basic course the parkway follows, was originally called Kararikong (place of the wild geese) by the Indians. Another spelling of the name is Karakung, which is derived from the name for the mortar or mill in which the Indians pounded their corn. This seems the more logical derivation because of the location there of grist mills. By 1643, the Swedes had transliterated the name to Carkoens. The English named the stream Mill Creek and, by 1683, it was called Cobb's Creek after William Cobb, who owned a grist mill along the banks.

College Avenue

Opened by affidavit in 1883, College Avenue was named for Girard College, which it bounds. Completed by 1848,

Girard College was built with $2 million left by Stephen Girard (see Girard Avenue), to care for the poor male orphans of the city. It was Girard's intention that every boy admitted be indentured to the city authorities until he was twenty-one; his entire care and education during his formative years would be assumed by the college. The site of the college was the farm of Peel Hall on the "Ridge Road." Girard's original idea was to have the school built on the block bounded by 11th, 12th, Market, and Chestnut Streets. That location became the first permanent home of the Community College of Philadelphia.

Collin's Lane

See Roxborough Avenue.

Columbia Avenue

See Cecil B. Moore Avenue, Counties.

Comanche Road

See Indian Tribes.

Comlyn Street

See Bringhurst Street.

Conrad's Lane

See Roxborough Avenue.

Consolidation

As settlers came to Penn's province and scattered beyond the limits of the little City of Philadelphia, they established villages and towns. As time went on, these villages and towns grew—and the need for services grew proportionately. Governments were formed by the people to provide for the needs of the inhabitants.

PHILADELPHIA IN 1854. The County of Philadelphia and the City of Philadelphia became one in 1854. The boroughs, districts and townships that were separate municipal entities were brought together by the wizardry of Eli Kirk Price. (Philadelphia City Planning Commission)

Consolidation

Between 1691 and 1854, twenty-nine separate governing districts were established within the confines of the County of Philadelphia. Prior to 1854, the City of Philadelphia was limited to the boundaries of Vine and South Streets, the Delaware and the Schuylkill rivers. Thinking to the future, public-spirited citizens, led by Eli Kirk Price (see Price Street), launched a drive in 1849 to consolidate all the smaller divisions into one gigantic city.

The Consolidation Act of 1854 effected this dream and created the new City of Philadelphia—contiguous with Philadelphia County. The act took away—on paper at least—many of the identifiable neighborhoods of Penn's time and later. Fortunately, Philadelphians are a difficult group to change. More than a century later, most of the old districts, boroughs, and townships are still referred to by the residents as neighborhoods. And even if slighted as neighborhoods, they are recalled by the names of streets, roads, avenues, even parks.

Prior to consolidation, the County of Philadelphia was composed of the City of Philadelphia; the districts of Southwark, Northern Liberties, Kensington, Spring Garden, Moyamensing, Penn, Richmond, West Philadelphia, and Belmont; the boroughs of Manayunk, Germantown, Frankford, White Hall, Bridesburg, and Aramingo; and the townships of Passyunk, Blockley, Germantown, Bristol, Oxford, Lower Dublin, Moreland, Byberry, Northern Liberties, Delaware, and Penn. One direct result of consolidating these many self-governing units was that a large number of continuous roadways had acquired different names as they passed from one area to another. To achieve some degree of order, four years after the city and districts became one, the city legislated to rename continuing streets and unify naming. Hundreds of streets had their names altered under that legislation.

In 1895, legislation was again introduced to alleviate the problem of name duplication. Again, hundreds of streets were changed—some going back to their original names, others assuming brand-new ones. Two years later, corrective measures were legislated to return historic names, lost through legislation, to their proper place in Philadelphia history.

Conwell Avenue

This avenue commemorates Russell Herman Conwell (1843–1925), lawyer, lecturer, and clergyman.

Following the Civil War, in which he served as a lieutenant-colonel, Conwell moved to Minneapolis, where he practiced law and founded the *Minneapolis Daily Chronicle*. Moving to Boston, he founded the *Somerville* (Mass.) *Journal*. While in the area, he became interested in a run-down Baptist church. After reviving it, he served as its minister. In 1882, he was called to repeat this success in Philadelphia at the Grace Baptist Church.

As a result of his efforts, the church prospered; he built a huge Baptist Temple, established three hospitals, and converted a night school he opened in 1884 into Temple College (now Temple University) in 1888. He served as Temple's first president. Conwell was one of the most popular speakers on the famed Chautaugua circuit. His lecture, "Acres of Diamonds," promoting the idea that opportunity is everywhere, earned Conwell over $8 million during his more than 6,000 individual presentations.

Cope Street

See Bonsall Street.

Corn Street

See American Street.

Counties

Pennsylvania counties are well represented in the naming of Philadelphia streets. Careful study of the information and material at hand suggests that a determined effort was made to name the streets to the north of

Counties

center city for counties in a manner similar to that used for the governor streets to the south (see Governors). The fact that the distance from Market Street is the same for both adds credence to this hypothesis: Mifflin Street, the first of the governors, is 1900 south; Berks, the first of the counties, is 1900 north.

But there the similarity ends. The governor streets succeed themselves in the same order as their human counterparts did in office. The county streets do not follow any set pattern, whether date of organization, separation by rivers, location as part of the disputed territories of Connecticut, Maryland, and Virginia, or similarity of name origin. The only pattern which does exist is that the counties are grouped together, separated by one of the rivers which flow through the counties outside of Philadelphia, Bucks, and Chester counties. I suspect that the original planner decided to start with Berks, named for the home county of the Penns, and end with Erie, the farthest county from Philadelphia. His successors probably sought to increase the grouping by adding Butler, Pike, Luzerne, and Lycoming. At any rate, the sequence as it presently stands is as follows.

Number of Blocks North of Market	Name of Street (County)	Date of Organization	Derivation of County Name
1900	Berks Street	1752	Berkshire, Eng.
2200	Susquehanna Avenue	1810	the river
2300	Dauphin Street	1785	son of Louis XVI
2400	York Street	1749	Duke of York
2500	Cumberland Street	1750	Cumberland, Eng.
2600	Huntingdon Street	1787	Countess of Huntingdon
2700	Lehigh Avenue	1812	the river
2800	Somerset Street	1795	Somersetshire, Eng.
2900	Cambria Street	1804	name for Wales
3000	Indiana Avenue	1803	the territory
3100	Clearfield Street	1804	"open fields"
3200	Allegheny Avenue	1788	the river
3300	Westmoreland Street	1773	Westmoreland, Eng.

Counties

Number of Blocks North of Market	Name of Street (County)	Date of Organization	Derivation of County Name
3500	Tioga Street	1804	the river
3600	Venango Street	1800	the river
3700	Erie Avenue	1800	the lake
3800	Butler Street	1800	General Richard Butler
3900	Pike Street	1814	General Zebulon Pike
4000	Luzerne Street	1786	Chevalier de La Luzerne
4100	Lycoming Street	1795	the creek

Other streets reputedly named for the Pennsylvania counties include Beaver, Bedford, Blair, Bradford, Cameron, Carbon, Centre, Chester, Clarion, Clinton, Columbia, Crawford, Fayette, Franklin, Fulton, Greene, Jefferson, Juniata, Lancaster, Lawrence, Lebanon, McKean, Mercer, Mifflin, Monroe, Montgomery, Northampton, Northumberland, Potter, Schuylkill, Snyder, Union, Warren, Washington, Wayne, and Wyoming. Of these, Beaver, Cameron, Franklin, Fulton, Jefferson, McKean, Mifflin, Monroe, Potter, Snyder, Washington, and Wayne were all men well known to Philadelphians, and it is possible that these names were given to the streets in honor of the men rather than the counties. The names of Chester, Lancaster, and Schuylkill are probably derived from their location or the direction in which they run. As for Greene, the Germantown street to carry that name was originally spelled Green. It was only when the city sought to remove similarly named streets that the final *e* was added.

Cox's Lane

See Manheim Street.

Crawford Street

See Counties.

Crefeld Street

Crefeld Street is named for a division of the old German Township, north of Somerhausen (now Chestnut Hill).

The first recorded deed for this street was for the road from Mermaid Lane to Moreland in 1887, at which time it was known as 28th Street. Roadbeds were acquired from then until the end of the nineteenth century. In 1900, 28th Street from Mermaid Lane to Springfield Avenue, Willow Grove to Abington, Gowen to Evergreen (vacated in 1959 and 1974), Hilltop to Rex (vacated in 1929), and Chestnut Hill to Norman Lane became Crefeld.

The village of Crefeld was named after a German town on the Rhine near Holland. In 1680, a colony from this region settled in Philadelphia. The Crefelders were noted as weavers of fine linen.

Crescentville Road

First confirmed as the Grubbtown Road in 1753, Crescentville Road extended from present-day Adams Avenue to Cheltenham. The town to which the road led was called Grubbtown as late as 1809. It was after the Crescent factory, located on the road to Jenkintown, was built and put into operation that the village's name changed.

Cresheim Road

Cresheim, a division of the old German Township, was settled by Mennonites, who came from Kriesheim (or Kriegsheim) in the Palatinate about 1683. The same name was given the creek, or stream, which formed the boundary between Chestnut Hill and Mount Airy. The word comes from the German *kriegsheim*, meaning war's home.

First confirmed in 1731 from today's Carpenter Lane to Germantown and Springfield, Cresheim Road extends from Westview Street to Cherokee. Additions to the original jury confirmation were made from 1899 to 1924.

Cresson Street

See Quarry Street.

Cross Street

See School House Lane.

Cullen Street

See Bradford Alley.

Cumberland Street

See Counties.

Curie Street

See Civic Center Boulevard.

CURTIN STREET. Andrew Gregg Curtin served as Pennsylvania's governor during the tumultuous Civil War years. His was one of very few governor portraits located in the Free Library of Philadelphia's Print and Picture collection.

Curtin Street

Curtin Street, first deeded to the city in 1916, has had as turbulent a career as its namesake, Andrew Gregg Curtin (1815–94).

Active in Republican politics from the age of twenty-five, Curtin campaigned for candidates in whom he believed—William Henry Harrison in 1840, Henry Clay in 1844, Zachary Taylor in 1848, Winfield Scott in 1852. As secretary of the commonwealth under Governor James Pollock (see Pollock Street), he secured larger appropriations for the public schools of Pennsylvania. In 1860, he was elected governor (see Governors). He was the first governor to be summoned to Washington by Lincoln to obtain support for filling state quotas for the

Union Army. Through his efforts, the Pennsylvania Reserve Corps was established. Known to the army as the "Soldier's Friend," Curtin established a fund for the support and education of war orphans. He took a leading role in the Altoona Conference of Union Governors, which established the solid position of their sentiments regarding the Emancipation Proclamation. He was re-elected in 1863, and then served as minister to Russia under President Grant after he was passed over as a possible vice-presidential candidate. When he supported Horace Greeley for president in 1872 on the Liberal Republican and Democratic tickets, he lost the backing of the Republican party. Changing to the Democratic side, he ran for Congress in 1881 and served three consecutive terms until his retirement in 1887.

Curtin Street, which extended from Broad to 20th Street in 1916, once traversed the city from Swanson Street to Penrose Avenue. The section between 7th and Swanson is now occupied by the bed of Packer Avenue. Most of Curtin Street has been stricken from the city plan and vacated, with just a block (Broad to 13th) remaining.

Custer Street

George Armstrong Custer (1839–76) graduated last in his 1861 class at West Point. Though court-martialed for dereliction of duty, Custer was mustered into the Union Army and commissioned in the cavalry. He quickly established a reputation as a dashing, daring and brilliant officer. He was brevetted brigadier general at 23; a year later, major general. At the end of the war, he was reverted to his permanent rank of captain.

In July 1866, Custer was appointed lieutenant-colonel of the newly-formed 7th Cavalry. He served as acting commander of that unit until his death.

Custer is best remembered for the Army's worst defeat in the Western campaigns: "Custer's Last Stand" at the Little Big Horn River, in present-day southern Montana.

CUSTER STREET. George Armstrong Custer was one of the youngest generals in the Civil War. A bold, aggressive leader, he met his match at Little Big Horn, Montana. (Matthew Brady Photo, National Archives)

Cypress Street

See Delancey Street.

D Dakota Street

See Indian Tribes.

Dallas Street

This street was named for George Mifflin Dallas (1792–1864), the only Philadelphian ever elected Vice-President of the United States.

A native of Philadelphia, Dallas was a staunch Democrat who held many state and federal elected and appointed positions. In 1844, he was nominated as James K. Polk's running mate. His most lasting accomplishment was that the Texas city of Dallas was named for him while he served as vice-president.

DALLAS STREET. George Mifflin Dallas was the only Philadelphian ever elected Vice President of the United States. He is also remembered by cities named Dallas in Pennsylvania and Texas.

Danen Hower's Lane

See Wister Street.

Darby Road

See Baltimore Avenue, Woodland Ave.

Darien Street

See Palumbo Plaza.

Darlington Road

According to gossip, Darlington Road was named to cure a bad case of homesickness. The road, from Kingsfield to Welsh Road, was dedicated in 1956. Streets department rumors say that one of the secretaries, usually quite competent, was slipping in her work because she missed her hometown in South Carolina. So to cheer her up, her co-workers arranged to have a new street named Darlington—for her birthplace. No one knows for sure if this action cured her illness.

Darrah Street

A Quaker lady, expelled from meeting, eavesdropped on the British and saved Washington's life. As her reward, a grateful city named a street in her honor.

Lydia Darragh (1728?–89) was inconvenienced by General Sir William Howe during his occupation of Philadelphia in the winter of 1777–78. In need of additional space, he commandeered the upper room of her house on 2nd Street, near Little Dock. One night, she secreted herself in a closet and listened to a British council. What she overheard chilled her heart: "The troops would march out on a certain night [late in the evening of December 4], attack Washington's army, and with their superior force and the unprepared condition of the enemy victory was certain."

Because of her patriotism, or love of her son Charles, who was with Washington, she told the British she needed a pass to buy flour at a Frankford mill. Changing her course of travel, she first warned the Americans. Backtracking to Frankford, she later obtained her sack of flour and returned home. The British marched back to Philadelphia from their raid "like a parcel of damned fools."

Lydia's tale was taught in the Philadelphia public school system for many years. It has since disappeared from the standard texts and is now considered a legend, but a school at 17th and Brown still bears her name.

Darrah Street (the spelling has been altered slightly over the years), from Bridge to Frankford, came into being as early as 1877. Presumably, this was the general location of the mill to which Lydia was supposed to have gone. The remaining section, from Meadow to Bridge, became Darrah in 1913, when the name of Cedar Street was changed.

DARRAH STREET. As tradition has it, Lydia Darrah learned of British plans to attack Washington's army, and proceeded to warn the patriots, thereby saving the rag-tag army and her son. Her tale was once taught in Philadelphia schools, but has since been relegated to the status of an old wives tale.

Dauphin Street

Named after Dauphin County, Pennsylvania (see Counties), Dauphin Street was confirmed by the road jury in 1853, from 30th Street to 33rd. Expansion, mostly by deed of dedication, extended the street to its full length by the time an 1877 jury confirmed its midsection, from 19th to old Lambs Tavern Road, roughly near 16th Street.

Dauphin County, once Paxtang Township in Lancaster County, was formed in 1785 and named in honor of Louis XVI's eldest son, Louis, the Dauphin. According to Espinshade, the word *dauphin* comes from the French province of Dauphiny, "so called from the dolphin assumed by the Courts of Viennois as a symbol of the mildness of their rule."

Davis Landing

See Fitzwater Street.

Deal Street

Frankford's first post office stood on Frankford Avenue where, if it went that far, Deal Street would cross. The first, and only, postman in Frankford in the early nineteenth century was John Deal. The post office was situated in the front room of his house. Tradition has it that Deal would receive the mail from the stage which traveled up Asylum Pike (now Adams Avenue), stuff the mail in his hat, and make his rounds.

Deal Street, from Kensington Avenue to just short of Leiper Street, was in use, at the very latest, in 1869. The small section to Leiper was added in 1892. The entire street was widened in 1907.

Dean Street

See Iseminger Street.

Dean's Alley

See Camac Street.

Decatur Street

Stephen Decatur (1779–1824), as a twenty-four-year-old Navy lieutenant, led a daring attack on the harbor of Tripoli to burn the *Philadelphia*, which had been captured by Tripolitan pirates. Only one member of his crew was injured in the foray.

Decatur is best remembered for his famous, but oft-misquoted, dinner toast: "Our country! In her intercourse with foreign nations may she always be in the right; but our country, right or wrong."

In 1824, Decatur, as a member of the Board of Navy Commissioners, was accused by Captain James Barron of conspiring to block Barron's promotion. The accusation resulted in a duel in which Decatur was killed. Stephen Decatur is buried in the graveyard at Philadelphia's St. Peter's Episcopal Church.

DECATUR STREET. In this famous painting of the Tripoli Wars, Philadelphia native Stephen Decatur is seen as the underdog. A gallant warrior, Decatur was killed in a duel.

Deer Lane

See "Things that Grow . . ."

De Kalb Street

De Kalb Street assumed the name of one of the bravest and most capable expatriates who served in Washington's army during the Revolution. Major General Johann Kalb (1721–80), son of a Bavarian peasant, volunteered his services and his experience as a major in the French army to Washington at Valley Forge. A thoroughly professional soldier, Baron de Kalb, as he was known in America, died as the result of wounds suffered at the disastrous Battle of Camden (South Carolina) in 1780.

De Kalb Street existed before the Civil War from Aspen Street to Fairmount. In 1885, a section, Locust to

DE KALB STREET. FAYETTE STREET. The Revolutionary War provided Philadelphia street namers with a seemingly endless list from which to draw. In this painting, we can see Baron de Kalb introducing the Marquis de Lafayette to Silas Deane. Deane was sort of the Oliver North of his day.

De Kalb Street

Walnut, was deeded to the city, but vacated in 1963. Another segment, from Cuthbert to Warren, became De Kalb when the name of Blodgett Street was changed in 1895. That too was vacated, in 1967.

Delancey Place

See Delancey Street.

Delancey Street

The Rt. Rev. William Heathcote DeLancey (1797–1865), a native of New York, was ordained an Episcopal priest in 1822. Subsequently, he was named assistant to Bishop William White of Philadelphia. A few years later, he resigned his clerical duties and was unanimously elected sixth provost of the University of Pennsylvania. In 1833, he became rector of St. Peter's Church at 3rd and Pine Streets. The back door of the house he lived in at 313 Pine opened onto what is now called Delancey Street.

The earliest section of Delancey Street, from Front to 4th, was called Union and predates the Biddle Directory of 1791. Before Philadelphia legislated to unify street names, the streets now called Delancey carried a variety of names: Powell (5th to 6th) was in use before 1805, Barclay (6th to 8th) before 1828, Asylum (Broad to 15th) before 1844, Rundy or Rundle (a dead end to 16th) before 1864, Talbot (38th to 39th) in 1883, and Woodland (39th to 40th) before 1862. Both Talbot and Woodland were stricken from the city plan in 1962. When new blocks were opened, after 1895, they were called DeLancey. The city changed the spelling of the name to Delancey in 1941.

Another section of Delancey Street, between 17th and 26th Streets, was changed from a Street to a Place in 1941. In 1895, this street had assumed Howell (19th to 20th), which was in existence by 1853; Walter (20th to 21st), in existence by 1856; and Cypress, in existence under that name since before 1897, earlier known as Trinity Place (22nd to 23rd), by 1877, and Factory (24th to 26th) since before 1862.

DeLancey was the first bishop of the Diocese of Western New York. And, in 1852, he became the first American bishop to assist in the consecration of an English prelate. During his tenure in New York, DeLancey helped organize Geneva (now Hobart) College.

Delaware Avenue

Prior to the late 1830s, the approach to Philadelphia's waterfront was alternately dusty and muddy—but always dirty, narrow, and irregular. The will of "mariner and merchant" Stephen Girard (see Girard Avenue) changed all that.

Girard had placed $500,000 of his estate in trust "to lay out, regulate, curb, light and pave a passage or street, on the east part of the city of Philadelphia, fronting the River Delaware, not less than twenty-one feet wide, and to be called Delaware Avenue, extending from Vine to Cedar [now South] street." His will also provided for improvements to Water Street. By 1839, a street fifty feet wide was built.

Girard's funding was unique; his idea was not. Merchant Paul Beck, Jr. (for whom Beck Street was named), like Girard, experienced the daily inconveniences of the little roadway to the docks; in 1820, he presented a plan to the council for a radical restoration. But no idea of such magnitude could be approved without first consulting Girard. And since Girard had an idea of his own, Beck's plan was never adopted.

Delaware Avenue grew to its present southern length between 1871 and 1919. The southern section was widened between then and 1928. The original segment was widened between 1858 and the present. The northern extensions were built between 1873 and 1971, and widened since 1909.

The name came from the river, which was erroneously named for William West, twelfth Baron De La Warr (1577–1618), who, the English thought, had discovered the river. In fact, Henry Hudson did.

In 1989, Philadelphia City Councilwoman Anna C. Verna presented an ordinance to change the name of this venerable street to Columbus Avenue. Her move came after a great deal of lobbying by various Italian-

DELAWARE AVENUE. Delaware Avenue drew its name from the Delaware River which flows nearby. The river was inaccurately named by the British who thought that Thomas West, Lord de la Warr, discovered the river. In truth, it was Henry Hudson. (Philadelphia Municipal Archives)

American groups. But the move was not without problems. Many native American groups, specifically descendents of the Delaware Indians, protested. Mayor Wilson Goode, in a Solomon-like act of politics, put the ordinance on hold until 1992, to see if the protests die down. Philadelphia's Traffic Division has been told to leave the Delaware Avenue signs up for two years, so the name change has not been made official by the Road Records Department. Chances are the City of Philadelphia, based on its record of honoring individuals with one-year renamings of streets (see Absalom Jones Way and Harry Hosier Way), will alter the name to Columbus Avenue for the 500th anniversary of Columbus' discovery of America—and then go back to the more traditional Delaware.

Devereaux Avenue

Devereaux Avenue, extending from the county line to the Delaware River, was named for Peter Devereaux, a "carrier" who lived at 47 Frankford Road in 1860.

The earliest record of Devereaux Avenue is dated 1878, from Castor to Oxford. A year later, the road jury confirmed another section, from Tulip to Tacony Street. The remainder, obtained by deed of dedication, was acquired in the twentieth century. The sections of the avenue between Tabor Road and Oxford Avenue were vacated in 1959.

Devon Street

See Bucknell Street.

Dewey Street

As does Bainbridge and Decatur, Dewey Street commemorates a naval hero: Admiral of the Navy George Dewey (1837–1917).

When Dewey received word of the outbreak of the Spanish-American War, he entered Manila Bay with his command of four cruisers and two gunboats. With his famous instructions to the captain of his flagship—"You may fire when you are ready, Gridley."—Dewey began the destruction of the Spanish fleet. With Dewey's action, the United States achieved the position of a major Pacific power. His cruiser, the *Olympia*, is berthed in Philadelphia along the Delaware.

Dickinson Street

Named for a man who voted against the Declaration of Independence—yet fought in the war for it—Dickinson Street first appeared as a single city block, from 4th Street to 5th, in the mid-nineteenth century. In 1853, the road jury confirmed that block. Though the court confirmed the action, a deed, dated 1848, indicates that the street existed prior to that ruling. The year after the first jury confirmation, Dickinson Street, in two separate Quarter Sessions Court rulings, was extended from east of 7th Street to the bulkhead line of the Delaware. By the end of the century, Dickinson was completed as far west as 33rd Street. By 1927, the street had reached its present length.

Though John Dickinson (1732–1808) voted against the Declaration of Independence, he and Thomas McKean (see McKean Street) were the only two congressmen who fought in the Revolution. Subsequently, Dickinson did support and sign the Constitution. He endowed Dickinson College in Carlisle, Pennsylvania.

Dicks Avenue

John Dick, a seedman, nurseryman, and florist, owned nurseries on a large section of what was then known as Kingsessing. Dick's nurseries were situated between 52nd and 53rd Streets on the west side of Darby Road (now Woodland Avenue) as early as 1857. Later, his holdings included the area between 63rd, 64th, and Lindbergh Boulevard, near Buist Avenue. The original Dicks Avenue ran through his lands.

The present configuration of Dicks Avenue is much

changed from that of the late nineteenth century. It now runs from Felton Street to Lindbergh Boulevard. Much relocated and redesigned, little or nothing of the original roadway remains.

Dock Street

"Dock street," according to the street directory of 1800, "is the only crooked street in Philadelphia, begins at the bridge in Front street; and extending northwestward in a serpentine track."

The street takes its name from the early creek whose course it follows. Before the English arrived, the stream was called Cooconocon. When Penn arrived in Philadelphia, the only public wharf was at the Blue Anchor Tavern (in the middle of present-day Front Street, about 146 feet north of Dock); the stream began to be called "the Dock."

Originally, Dock Street was on either side of the main branch of the creek, between the Delaware River and 3rd Street. Though opened by affidavit in 1883, Dock Street was built in 1784 when the entire creek was arched over. The "serpentine" shape of the street is due to the course of the stream.

Drexel Road

Drexel Road is named for the Drexel family, whose members include Anthony Joseph (1826–93), philanthropist, banker, and founder of Drexel University, and Francis Martin (1792–1863) and Joseph William (1833–88), both bankers.

Drinker's Alley

See Quarry Street.

Drinkwater Street

See Keyser Street.

Duane Street

See Camac Street.

Dubree Street

See Cameron Street.

Duffield Street

A pathway which led from a dam on Little Tacony Creek evolved in the nineteenth century into a street commemorating the dam-builder.

Duffield's Dam fed a small tributary of the creek into Waln Grove, the country manor of Robert Waln (see Waln Street). The small dirt path was later improved, covering the small creek, and became Willow Street about 1863. Willow, named for the abundance of such trees along the creek, became Duffield in 1913. Other sections of the street, from Bridge to Brill and from Comly to Frankford Avenue, were included in the 1920s. Though the dam and Waln Grove are gone, Duffield Street still leads to a beautiful setting—Cedar Hill Cemetery and Wissinoming Park.

Over the years, folk historians have listed Benjamin Duffield, who met Penn at the dock in 1682, Benjamin's son Edward, a pioneer watchmaker, and the Rev. George Duffield, first pastor of Old Pine Presbyterian, as donors of the name.

Dunks Ferry Road

On December 12, 1776, George Washington ordered that Dunk's Ferry be carefully guarded. When he was ready to cross the Delaware on his surprise visit to the Hessians at Trenton, supporting troops under General John Cadwalader were to cross at Dunk's Ferry. But "the river was so full of ice, that it was impossible to get the artillery over." Thus, Washington's plans were foiled and Dunk's Ferry seems to be all but forgotten, except for the road which bears its name.

Dunks Ferry Road once led directly to the Delaware River—to the old Dunk's Ferry Hotel, which dated back to 1733. Now, however, the road extends only from Byberry Road to the county line. The ferry itself was probably in existence by 1710, when it was confirmed by Quarter Sessions Court.

Dunlap Street

This street was named for John Dunlap (1747–1812), printer and publisher of *The Pennsylvania Packet, or The General Advertiser*. The Declaration of Independence was first published in his office from Jefferson's manuscript.

Durass Street

See Fulton Street.

Duy's Lane

See Wister Street.

Dyott Street

In 1896, plans were approved for "a new street . . . to be called Dyott Street." It was to extend from Beach Street to an angle northeast of the Aramingo Canal. Three years later, it was opened by ordinance from Beach Street to Norris. The roadbed, formerly a part of the Aramingo Canal and later Aramingo Avenue, was the general location of the village of Dr. Thomas W. Dyott.

Dyott Street

Dyott opened the Dyottville Glass-Works in 1833. To accommodate his large work force, Dyott (whose medical degree might have been manufactured) established four hundred acres of Kensington as Dyottville, complete with farming, a chapel, and a temperance society. From his glassworks, Dyott expanded to banking. Unfortunately, his expedition into finance cost him three years in prison for "fraudulent insolvency."

Eastwick Avenue

Though city records do not reflect the dating of the original Eastwick Avenue, it is certain that an Eastwick Avenue existed as early as 1858, when Elizabeth's Avenue, "south from Ogden above 16th," became known by that name.

Present-day Eastwick, bearing the same name as the section of southwest Philadelphia through which it runs, is recorded by deed of 1890 from 89th Street to former 90th. Extending today from 58th Street to Lindbergh Boulevard, the avenue was mainly developed in the early to mid-twentieth century.

Andrew Eastwick (1811–79) was a partner in the locomotive-building firm of Garrett & Eastwick. In 1843, trading under the name of Harrison, Winans & Eastwick, the firm began construction of the Russian railroad from St. Petersburg to Moscow. In 1850, Eastwick acquired Bartram's Gardens (see Bartram Avenue) and erected a mansion for himself, which was destroyed by fire in 1896.

Edgley Avenue

Edgley (see Orion Road), one of the estates purchased for inclusion in Fairmount Park, was from 1826 to 1836 the summer residence of Dr. Philip Syng Physick. The name for the avenue was given in 1895, when Midvale—40th to 49th Street—underwent a name change. Ironically, that is the only section of Edgley which remains today. Other segments—Windemere to Steinberg Avenue—were vacated in 1955–56.

Edison Avenue

In the Philadelphia tradition of naming streets and roads in honor of national figures, Edison Avenue was named for Thomas Alva Edison (1847–1931). Edison, who had only three months of formal education, invented such things as the incandescent light bulb, the phonograph, the "kinetograph" camera and the "kinetoscope"—

forerunners of today's film cameras and video camcorders. During his lifetime, he was granted more than a thousand patents, and became the personification of the practical American genius. Edison once described genius as being "one percent inspiration and ninety-nine percent perspiration."

Elberon Avenue

Running from Rhawn Street to Hoffnagle, Elberon Avenue takes its name from the section of Philadelphia through which it runs. The first section to be deeded to the city was in 1886 from Solly Avenue to Stanwood. The section from Rhawn to Stanwood, in use by 1874, was officially opened in 1895.

Elbow Lane

This little street, running from 3rd to Bank Street, resembles an elbow on early city maps. Opened by affidavit in 1883, it was in use during the early eighteenth century. In 1895, its name was changed to Ludlow Street; two years later, it reverted to Elbow Lane. (See Bank Street.)

Eleanor Street

See Bancroft Street.

Elfreths Alley

Philadelphia's most famous alley obtained its name by virtue of a marriage.

John Gilbert, an early arrival from England, settled in the Byberry area. He also had a house on a small alley which ran from Front to 2nd Street. In the early eighteenth century, Gilbert's eldest daughter married Henry Elfreth. Besides winning Sarah's hand, Henry got the alley.

ELFRETH'S ALLEY. The daughter of early settler John Gilbert married Henry Elfreth and, as a wedding present, received a small house on a small alley, running from Front to 2nd Street. Elfreth's Alley is reputed to be the oldest continuously occupied street in Philadelphia.

Elfreths Alley

Reputed to be the oldest continuously occupied street in Philadelphia, Elfreth's Alley became a street in 1865. Thirty years later, it was added to Cherry Street. In 1937, because of pressure from civic and historical groups, the little street regained its original name.

Eliza Street

See Ingersoll Street.

Elizabeth's Avenue

See Eastwick Avenue.

Elmwood Avenue

Elmwood Avenue—first deeded to the city in 1870, from 58th Street to Hay Lane (the line of modern 62nd Street)—commemorates an early village in the vicinity of 89th Street. The first major growth of this avenue was in 1875, when the road jury confirmed it from 58th Street to Island Avenue. By 1925, Elmwood Avenue was complete to its present distance from Lindbergh Boulevard to Cobbs Creek Parkway.

The name itself is probably derived from the numerous elm trees which were found in the woods of that area.

Emma Street

See Water Street.

Engleside Street

See Bailey Street.

Erie Avenue

See Counties.

Essington Avenue

Essington, the quarantine site during the yellow fever epidemics of the eighteenth century, was located on Tinicum Island. Essington Avenue was named for that location.

The records of this avenue are quite confusing. According to the city's official legal status of streets, Essington Avenue was built to its present configuration between 1931 and 1943. However, the city solicitor indicated in 1937 that, in the section from 90th to 92nd Street, the roadway had been in use for "at least twenty-one years." In addition, an ordinance was passed in 1915 to widen the avenue between 82nd and 84th Streets. It is probable that the road was begun shortly after Tinicum Island, and consequently Essington, was linked to the mainland in the late nineteenth century.

Evangelist Street

See Fulton Street.

Factory Road

See Axe Factory Road.

Factory Street

See Delancey Street.

Fairmount Avenue

On Holme's first map of the "Province of Pennsylvania," a prominent hill is sketched in to the north of Penn's city, not far from the banks of the Schuylkill. On this 1681 chart, it is designated "Faire Mount." "Fare Mount," a 1710 visitor wrote, "is a charming spot, shaded with trees." That charming spot ultimately became the 4,077.59 acres of Fairmount Park.

Fairmount Park began in 1812, when Philadelphia purchased Morris's Hill, Faire Mount on Holme's map, as the site of the city's waterworks and reservoir. The reservoir area is now occupied by the Philadelphia Museum of Art.

The avenue which leads to the lower entrance to the park is Fairmount Avenue. The earliest section of this roadway, from the Delaware River to Front Street, was deeded to the city by Thomas Coates in 1771. Coates's street was opened from the Delaware to Ridge Avenue the next year. Within fifty years, Coates extended as far west as the Schuylkill. By the end of the nineteenth century, Fairmount Avenue terminated at present-day Haverford Avenue. Old Coates Street became Fairmount in 1873, between the Delaware and Schuylkill rivers.

Other names for this avenue included Hickory Lane, from Old York Road to Ridge Avenue; Vineyard Lane; and Plumstead Lane, which had been Francis Lane and then New Hickory Lane, from Ridge Avenue to the Schuylkill.

Fall's Lane

See Queen Lane.

Farragut Street

From the time of its earliest dedication in 1870, from Chester Avenue to Springfield, this street honored David Glasgow Farragut, the American naval hero.

Farragut (1801–70) entered the United States Navy as a midshipman at the age of nine. As a teenager, he saw service during the War of 1812 and took part in action against the incursion of pirates and freebooters off the southern coast of Florida. During the Civil War, he commanded a fleet on the Mississippi which captured Fort Jackson, New Orleans, Baton Rouge, and Natchez. His most famous victory was Mobile Bay. As an admiral, Farragut was promoted as a running mate for U.S. Grant in the 1864 presidential election. Neither ran.

Farragut Street, from Baltimore Avenue to Chester, became known as Markoe to commemorate Abraham Markoe, a founder of the First City Troop, in 1895. Three years later, that section was changed to Farragut Terrace. The northernmost lengths, from Market to Spruce, were deeded to the city between 1906 and 1911 as Markoe Street. The section of Markoe from Spruce to Walnut became Farragut Street in 1910; the section from Walnut to Market, in 1943.

Farragut Terrace

See Farragut Street.

Faunce Street

Present-day Faunce Street, from Central Avenue to Roosevelt Boulevard, is a product of twentieth-century planning. The first section to bear this name, from Morton Street to the Northeast Freeway, dates to 1915. The last, the segment from Bradford Street to the

Faunce Street

Boulevard, was approved in 1957.

Situated in the Fox Chase section of Philadelphia, a Faunce Street existed on the city plan as early as 1887. The Faunce family was numerous in that area; in 1860, there were more than forty family members listed in the city directory. Most of them were fishermen from the Richmond district. Two, George W. and William, were shipwrights. Their main ancestor, John, came to America in 1623 and settled in New England. His son Thomas was a ruling elder at Plymouth.

Fayette Street

See Counties.

Federal Street

In 1790, the commissioners of the District of Southwark established Federal Street as a "new street," from the Passyunk Road to the Schuylkill River. Eleven years later, the governor ordered the street to be opened from Swanson Street to the Schuylkill. It does not, however, appear on published street lists until 1813. The section of the street from Front Street to the Delaware River was stricken from the city plan in 1914.

The name Federal was given to the street because it led from the early navy yard to the federal arsenal near Grays Ferry.

Fetter's Lane

See Cherry Street.

Fillmore Street

This street is named for the thirteenth president of the United States, Millard Fillmore (1800–74). (See Bouvier Street.)

"Filthy-dirty"

Are dirty, pothole-ridden streets symptomatic of the decline of modern civilization? Or are Philadelphians unknowingly perpetuating a peculiar vestige of the city's rich historic past?

"Filthy-dirty" has been one of Philadelphia's pseudonyms for almost three centuries—at least. In fact, Robert Venable, a man with a remarkable memory born in 1736, recalled hearing that descriptive title used in his youth. As Venable remembered them, Philadelphia streets were alternately muddy or dusty, depending on the weather. A citizen took his or her life in hand traversing the city, dodging garbage and other nauseating obstacles. Sidewalks, the protective corridors we know today, existed only in rare instances.

Some private homeowners, at their own expense, installed brick or fieldstone walkways from the houseline ten feet into the rut-filled cartways. Bricks were the first choice. But the demand quickly exhausted the supply . . . and skyrocketed the cost. When brick was priced out of reach of the common man, Philadelphians filled the remaining spaces with pebbles fished from creek and river beds.

As late as 1718, the provincial government could not bring itself to accept the responsibility for maintaining the streets of the city. As a result, many residents took matters into their own hands. They could stand the filth and potholes no longer. They paved the streets themselves . . . "from ye Kennel [gutter] to the middle of the streets before their respective tenements with pebblestones." However, not every homeowner was filled with civic pride. Some felt they were taxed enough to cover the cost of surfacing the roads. Let the city do it, they cried!

Other citizens were perturbed by stray horses and rumbling wagons accidentally destroying their fences and property. To protect their homesteads, they installed short, stout posts at regular intervals along the street edge of their properties. Outside these posts, "the middle part of the streets [was] very dirty, [and] interrupted frequently by various types of lumber."

The Pennsylvania grand jury recognized the problem in 1750, and reported that "frequent complaints

"Filthy-dirty"

[were] made by strangers and others of the extreme dirtiness of the streets for want of paving." But the knowledge of the severity of the problem didn't trigger any immediate response—or action. Eleven years later, a lottery was held—probably at the instigation of Benjamin Franklin—to raise enough money to pave some Philadelphia streets. The response to the lottery was by no means overwhelming.

City fathers raised enough money—$7,500—to pave North 2nd Street, from High (now Market) to Race. The work was begun by unskilled laborers. Their craftsmanship was so poor that a British soldier, and skilled paver, John Purdon, was given leave from the army to supervise the work. Purdon's roadway, by modern standards, left much to be desired. He installed the largest stones he could locate in the middle of the road; smaller stones were placed to either side. His planning provided the streets with excellent drainage, but wreaked havoc on the springs and sprockets of the crudely-constructed carriages and wagons—and the *derrieres* of the passengers.

Despite the installation of some paved streets, Philadelphia remained "filthy-dirty." Laws were enacted that required residents to sweep the sidewalks in front of their properties every Friday. Where did they put the refuse? Where else? In the street! By 1765, the city assumed its responsibility for the area between the sidewalks. In that year, Robert Erwin was appointed city "scavenger." His job was to clean all the streets—once a week. Philadelphia's streets did not improve, even with Erwin's efforts.

By the end of the Revolution, few streets were paved. In 1783, homeowners on Lombard Street, between 3rd and 4th, petitioned the Board of Street Commissioners for paving. They explained they had "cheerfully paid their portion of the street taxes, in full confidence, that as soon as the situation of our public affairs would admit, they should be relieved in the premises." They were angry because, as they wrote, all of the other east-west streets had been paved westward to 5th Street. When all petitions and citizen demands failed, some homeowners threw up their hands in disgust; others acted. After one of the Whartons was thrown from his horse—it had stepped in a pothole—on 2nd Street, between High Street and Chestnut, his friends raised the money to pave the street. William Sansom, trying to make his row of houses more saleable, offered to advance the joint city councils the necessary money to pave two additional blocks of Walnut Street to 8th. The city didn't take him up on his offer, so Sansom paid for the work himself. He did get even; he named his privately-funded street "Sansom" after himself.

Times have changed. But have they? Philadelphia still has filth and dirt on its streets. Vehicles are still going into the shop for new shock absorbers. Maybe citizens should resign themselves to the condition and place historical markers on the dirtiest streets.

First Names

Because the city lacks definitive information on the naming of streets, it is an almost impossible task to determine the sources for those streets which bear someone's first name. There are almost two hundred and fifty such streets in Philadelphia, ranging from Abigail to Zeralda. Those few—like Catharine and Wigard—for which I could find a likely donor are listed separately.

Fishers Lane

Fishers Lane, a small street that once extended from Germantown Avenue to Wyoming Avenue, commemorates the Fisher family of Philadelphia. The most illustrious member of that family was Joshua, who in 1756 completed, for "the Merchants and Insurers of the City of Philadelphia, This Chart of Delaware Bay and River, containing a full and exact description of the shores, Creeks, Harbours, Soundings, Shoals & Bearings." Fisher's was the standard map until the United States Coast Survey in 1846, though it was suppressed immediately following its completion for fear it would fall into French hands.

The lane, of which little remains, was confirmed in 1785. Portions of it disappeared into the beds of East Logan and Ruscomb Streets.

Fishtown

See Kensington Avenue.

Fitzwater Street

The name of this street was the result of a clerk's typographical error in the early nineteenth century. Confirmed in 1798 from 7th Street to the Passyunk Road, the street was named Fitzwalter. In the oldest mounted plans of the District of Moyamensing in the streets department files, this name appears. Yet a layout of the streets in Moyamensing, dated 1829, spells it Fitzwater. Apparently, in transcribing the street names to an updated map, the clerk inadvertently omitted the *l*.

The original spelling of the name commemorated Thomas Fitzwalter, who came to this country aboard the *Welcome* with William Penn. On the voyage to the new world, Fitzwalter lost his wife and two of his four children. He served as a member of the Pennsylvania Assembly in 1683, representing Bucks County.

In 1895, to make way for an unbroken Fitzwater Street, the council changed the names of German Street (Passyunk Avenue to 2nd Street), Mead (2nd to Swanson), and Davis Landing (Swanson to Delaware Avenue). Complete to its present length by 1928, Fitzwater now extends from Grays Ferry Avenue to Delaware Avenue.

Flat Rock Road

A "conspicuous body of rocks" filled a narrow valley in the upper reaches of Manayunk. As a result, the valley was commonly called Flat Rock, as was the village that was later known as Manayunk (see Manayunk Avenue). The bridge across the canal bears the same name. And so did the woolen and cotton mills which were established in that locale in the nineteenth century.

As early as 1864, an avenue by that name is listed "from Columbia Bridge W Canal, Myk." The road jury confirmed Main or Nixon Street, from Canal to Fountain

Street, in 1822. However, it appears the road was not started immediately.

Though the official name for this road was Main or Nixon, it seems likely that it was called Flat Rock by the local residents. The common name was made legal in 1921. The full length of Flat Rock Road was achieved by 1873, from Leverington Avenue to northwest of Fountain Street.

Fletcher Street

Very few people live to see their names emblazoned on a street sign, because the honor is usually bestowed posthumously. Joshua S. Fletcher was one exception. He and his son, Joshua Jr., were living at 2813 Fletcher Street in 1880, when the street made the official index of the Philadelphia street directory. Fletcher worked as a "coach trimmer" in 1840, an editor in 1859, and a "receiver" in 1860. Though the street's name was not official, he was listed in 1870 as living on "Fletcher, near 28th street."

The earliest street to be called Fletcher was the block on which he lived; it probably dates to before 1854. Other sections were included and the streets widened by 1895. In that same year, such named streets as Moore (Emerald to Amber), Townsend (Sepviva to Cedar), Capewell (Gaul to Belgrade), and Ash (Thompson to Delaware Avenue) had their names changed to Fletcher.

Florence Avenue

Thomas Birch Florence (1812–75), a native of Philadelphia's Southwark, was a noted figure of the nineteenth century. After his hat business failed in 1841, the politically active Florence was elected secretary of the board of Philadelphia's public schools. He must have served with some distinction because a school, once at 8th and Catharine Streets, was named in his honor. A park has since replaced the school—but the Florence name remains.

The West Philadelphia Homestead Association, headed by Florence, developed the area through which this avenue runs. The street was dedicated in 1859, while Florence was serving as a member of the U.S. House of Representatives. The original avenue extended only from 49th Street to the Philadelphia & West Chester Railroad. Additional segments, from 48th to Cobbs Creek Parkway, appeared from 1893 to 1916.

Florence Street

See Bouvier Street.

Ford Road

Adopting its name from an old trail leading to an eighteenth-century crossing over the Schuylkill, Ford Road returned to the official Philadelphia records by 1878.

The original road existed on the west side of the river, almost opposite Strawberry Mansion, and led to Robin Hood Ford, which was at a point directly south of Laurel Hill Cemetery. The tavern at the ford, with its signboard of Robin Hood, also lent its name to Philadelphia's natural amphitheatre in Fairmount Park, Robin Hood Dell.

Present-day Ford Road was enlarged to its present width from 1902 to 1957.

Fort Mifflin Road

See Mifflin Street.

Fox Chase Road

Fox Chase was a village in the old Dublin Township prior to the 1854 consolidation (see Consolidation). The road which commemorates the village runs from Algon Avenue to Castor.

The community of Fox Chase took its name from

the sign board of a late eighteenth-century inn, located at Asylum Road (now Adams Avenue) and Olney Avenue. The sign was "a picture of mounted huntsmen in red coats, and Nathan Hicks, the proprietor, holding the foxes."

Another suggestion is that Tobias Leech (see Ashmead Street), a settler in the area during Penn's time, was known for his hunting grounds, which were kept up by his descendents, and this originated the name.

Legal recordings of Fox Chase Road indicate it was deeded to the city along its full length between 1948 and 1951.

FORT MIFFLIN ROAD. Fort Mifflin stood as the final defense of the Delaware River in 1777. Together with its sister-fort, Mercer, on the Jersey side of the river, Mifflin held off the British fleet long enough for Washington's troops to retreat to Valley Forge. Fort Mifflin was originally on Mud Island, but time and tides have brought the fort to the shore. (Photograph by Jack E. Boucher, Historic American Buildings Survey)

Fox Street

See Water Street.

Frankford Avenue

Taking its name from the section of Philadelphia through which it runs, Frankford Avenue was first confirmed from Front and Vine Streets to the Poquessing Creek in 1747, though the road appeared on maps as early as 1680. In 1795, it was extended from the Frankford Creek to the Pennypack Creek. By the time of the 1854 consolidation, the avenue was confirmed from the District of Kensington to the Borough of Frankford. Combined with the other jury confirmations, Frankford Avenue was complete to the distance it extends today. Once part of Philadelphia's turnpike system (see Toll Roads), and known as the Frankford and Bristol Turnpike, the avenue was freed from toll in 1892.

Though some folk historians infer that the section of Frankford was named after a man called Frank who operated a ford or ferry in the neighborhood, it is more likely derived from the Frankfurt Company, which owned land in the area in the time of William Penn.

FRANKLIN SQUARE, FRANKLIN STREET, ETC. The undisputed "man for all seasons" must be Benjamin Franklin. He was involved in the drafting of the Declaration of Independence, the harnessing of lightning, the development of street cleaning and street lighting, the creation of fire companies and fire insurance companies, and colleges and schools for the working class. (Painting by Joseph Duplessis, National Archives)

Franklin Square

See Squares.

Franklin Street

See Brill Street, Girard Avenue, Counties.

Frazier Street

See Alden Street.

Front Street

The first street of Penn's city, or the one which fronted on the Delaware River, was surveyed for building lots as early as 1682, on the west side of the street, from Walnut to Chestnut. The entire street, both sides, had been surveyed from Vine to Cedar (now South), with few exceptions, by the beginning of the eighteenth century.

In 1755, Thomas Pownall wrote: "Front Street stretches farther along the banks of the Delaware than as designed by the original plan, as the other streets are more and more defalcated of their length, so that the shape of the town at present is that of a semi-oval." Though surveyed and in use, the street was not officially listed until 1804.

The section of Front Street which travels through the old District of Southwark was confirmed in 1790 from present-day South Street to Hoffman Street. This section was ordered to be opened April 6, 1819, under an act of 1787, "confirmed by the Governor and Supreme Executive Council."

The northern stretches of Front Street were in public use, at least as far as Hunting Park Avenue, before the end of the nineteenth century. The remainder, from Hunting Park to 65th Avenue North, was complete by 1961.

At one time Philadelphia had two Front Streets—Front Street Delaware and Front Street Schuylkill (see Original Streets).

Fulton Street

It seems likely that in 1895, when the city sought to name streets uniformly, it would seek to provide a nautical name for the street that led to the U.S. Naval Asylum. Fulton Street, which runs from Taney Street to 2nd, honoring Robert Fulton, would fit that bill.

Fulton (1765–1815) invented the first sailing vessel to be propelled by steam. The *Clermont* was launched on the Hudson River in 1807. Fulton had worked as a goldsmith in Philadelphia before going to New York.

The street, under several other names, dates back as early as 1836, when it ran from 2nd Street to 3rd and was known as Concord. Other names include Naval Asylum Place (Taney Street to the Naval Home, since vacated), McCrea (Park Avenue to Juniper, since vacated), Evangelist and St. Paul (8th Street to a dead end), Durass (the dead end to 6th Street), Campbell (Randolph to 6th), and Harmony (5th Street east to a dead end). All these streets became known as Fulton in 1895.

Robert Fulton was also honored by his adopted state with the naming of Fulton County in 1850. Originally recommending the name Liberty County, the bill was amended to honor the inventor.

FULTON STREET. Though Robert Fulton had no strong ties to this particular street, it appears that in 1895, when Philadelphia sought uniformity in street naming, they picked a "nautical" name for the road to the U. S. Naval Asylum. Before moving to New York and making his fame, Fulton worked as a goldsmith in Philadelphia. (Scharf and Westcott's *History of Philadelphia, 1609–1884*)

Galloway Street

Though records do not exist to certify the assumption, it is possible that Hermitage Street, from Green to Fairmount Avenue, viewed by the road jury and confirmed by Quarter Sessions Court in 1810, was on land once owned by Joseph Galloway. An eminent lawyer and a staunch Loyalist, Galloway was "held an authority in all matters touching real estate." An intimate friend of Benjamin Franklin, he was the custodian of the elder statesman's valuable papers and letter-books when Franklin went to England in 1764. In 1776, Galloway joined Howe and, during the British occupation of Philadelphia in 1777–78, served as superintendent of police and of the port. After the British left, all of Galloway's property, including Ormiston, his mansion near Laurel Hill, was confiscated.

The first recording of a street named Galloway was in 1858, from George to Cambridge. In 1895, Hermitage became Galloway. Other sections of this street, now much retracted, were deeded to the city between 1914 and 1957.

Gallow's Lane

See Schuylkill Avenue.

Garden Street

See Blair Street.

Garfield Street

James Abram Garfield (1831–81) was twentieth president of the United States.

A native of Ohio, Garfield served as a Republican member of that state's senate. Following the outbreak of the Civil War, Garfield recruited a volunteer regiment and became its colonel. As a result of outstanding performance at Shiloh and Chickamauga, he was promoted to major general of volunteers. He resigned his commission in 1863 to represent Ohio in the U. S. Congress. He was elected and served with little opposition until 1874, when his involvement in the Credit Mobilier scandal was disclosed. In 1880, Garfield was elected to the Senate, but never served. He was chosen as a compromise candidate for the Republican presidential nomination. Along with Chester A. Arthur, Garfield won the election with a small popular majority.

In 1881, while waiting in a Washington railway station, Garfield was shot by a disappointed office-seeker, Charles J. Guiteau. Garfield lingered for 11 weeks, during which time the question of whether Arthur could become president or acting president was hotly and publicly debated.

Garside Street

See Bonsall Street.

Gates Street

Gates Street is another Philadelphia roadway commemorating a Continental Army general. Horatio Gates (1728/29–1806) won the 1777 battle of Saratoga with the brilliant help of Benedict Arnold. Inflated with his own grandeur, Gates did nothing to stop the machinations of the Conway Cabal, an attempt to replace Washington with Gates. His Saratoga victory was overshadowed by his disastrous defeat at Camden, South Carolina, in 1780.

Geary Street

See Governors.

George Street

See Sansom Street.

Gerhard Street

See Sartain Street.

Germantown Avenue

In 1693, when the first settlers traveled to their new homes in what is now Germantown, they took a strange, winding path through the woods. The Indians had blazed this trail and, by habit, had selected the easiest, if not the most direct, route. Gradually, their footpath became a road, following much the same twists and turns as today's Germantown Avenue.

The early immigrants from Crefeld, Germany (see Crefeld Street), built their town along what they called Main Street. For companionship, as well as mutual protection, they built their houses in two rows, facing each other across the rough trail. The backs of the homes opened onto their farmland. The leader of the Crefelders—Francis Daniel Pastorius, for whom Pastorius Street was named—wrote: "The path to Germantown has by frequent going to and from been so strongly beaten that a road has been formed."

In March 1709, the inhabitants "Humbly" petitioned the assembly "that your Peticoners Haveing Plantacons lying Very Remote in the Country and In the Edge or Outskirts of this County, And It being Very Difficult for them to pass and Repass their Said Plantacons by Reason there is No Publick Road Laid out far enough to Reach to the said Plantacons . . . [the assembly should] Allot Some Convenient Plans for Laying Out a Road from the Late House of Edward Lane Deceased being on the Queen's Highway unto Mannitania." The "road" to Germantown, improving the earlier Indian path, was built several years later. Governor William Pownall attested to this fact in 1754, when he traveled "another great road, which goes from Philadelphia . . . to [Harris's] ferry, but keeps to the N. E. side of the Schuylkill, and runs through Germantown, &c. to Reading."

Though a road now existed, it was insufficient to handle the large volume of traffic. Rain churned the dusty road into a quagmire of mud. "In the spring of the year, especially, the way was only possible with the greatest of difficulty." Wagonmasters constantly complained of their horses' spraining and breaking legs in the ruts and the mud. "At certain periods of the year, there was non-intercourse between Philadelphia and Germantown."

In March 1802, "the President, Managers, and Company of the Germantown and Reading Turnpike Road" were incorporated to build a road through Germantown to the top of Chestnut Hill, through Hickorytown, the Trappe, and Pottstown, to Reading. The company "macadamized" the road and completed the task by 1804.

The Germantown Road became an avenue in 1858. It increased in size in 1931, when the old Chestnut Hill and Springhouse Turnpike and the Perkiomen Turnpike were added. The last section of Germantown Avenue was opened in 1914.

Gillingham Street

Gillingham Street was originally a small road leading to a dam on Little Tacony Creek in Frankford. Joseph J. Gillingham built and maintained the dam, which was reached by what was then called Ridge Street.

Opened by affidavit, Ridge Street was actually in use by 1862, if not earlier. In 1895, it became Gillingham. Other segments of the street were added in the late nineteenth and early twentieth centuries.

Girard Avenue

Very few men have done so much for Philadelphia—or have been more misunderstood—than Stephen Girard (1750–1831). Girard College, Delaware Avenue, Girard Bank (now Mellon Bank), the list of contributions Girard made to his adopted city is practically endless.

The first Girard Avenue, confirmed in 1845, was from Ridge Avenue to Corinthian, next to Girard College (see College Avenue). In 1858, Girard Avenue grew with the inclusion of old Franklin Street (from 6th to Frank-

Girard Avenue

GIRARD AVENUE. No one man has done more for the City of Philadelphia than Stephen Girard. The financier established one of the city's most powerful banks, a school for orphaned boys, Girard College, and came up with money to improve Delaware Avenue.

ford Avenue) and Prince (from Frankford to Norris). West of the college the avenue grew in spurts: in 1856, to 33rd Street; in 1868, to 64th Street; and, by 1936, as far west as 68th.

For one year, beginning July 29, 1982, a part of Girard Avenue, running between 34th Street and Belmont Avenue, was renamed John Gloucester Way. This was done to commemorate the 175th anniversary of the First African United Presbyterian Church. First African was founded in 1807 by John Gloucester and was the first black Presbyterian congregation organized in the United States. It was used as a way station in the Underground Railroad during the days of slavery. At the end of the anniversary celebration, that section of street "reverted to its previous designation."

Godfrey Avenue

While visiting Stenton, Thomas Godfrey, a Philadelphia glazier, noticed a piece of broken glass reflecting the sun. After consulting a book by Isaac Newton, which he found in James Logan's vast library, he constructed a mariner's quadrant, which was used by Joshua Fisher (see Fisher's Lane) to chart the Delaware Bay. But Godfrey's idea was pirated; today the instrument, with little or no change, is called Hadley's Sextant.

Godfrey is, however, remembered in the name of an avenue. The earliest record for a Godfrey Avenue dates to 1858, when Harrison Avenue, from 20th Street to Wister, became Godfrey. There is no official date for Harrison, but it was listed as a street in use by 1844. Wakeling Street, from Rising Sun Avenue to Rutland Street, became Godfrey in 1895. The rest, from Castor Avenue to Woodlawn Street, was in use by 1951.

Gold Street

See Brandywine Street, Moravian Street.

Gorgas Lane

Overlooking present-day Wissahickon Avenue is a three-story stone building, the "Monastery of the Wissahickon" (see Monastery Avenue), the citadel of the Seventh-Day Baptists, who were led by Joseph Gorgas. Gorgas came to Philadelphia from Lancaster in 1745 and settled on an eighty-acre farm, now part of Germantown. In 1761, the Monastery was sold to Edward Milner. Three years later, the court confirmed the road from Ridge Avenue to the Wissahickon. This was the beginning of Gorgas Lane.

In 1811, the roadway, from Germantown Avenue to present-day Stenton Avenue, was confirmed. The section between Cresheim and Mower was included in 1900; the section from Wenthan to Cresheim, in 1925, by the addition of old Hepburn. Gorgas Lane, from Stenton to Cheltenham Avenue, was deeded to the city in 1928. The lane reaches only to Michener Avenue today; the last segment to be added, Pickering Avenue to Cheltenham, was stricken from the city plan in 1947.

Gorgas Mill Road

See Mount Airy Avenue.

Gothic Street

See Sansom Street.

Gould Street

GOULD STREET. The top financial rascal of his day was Jay Gould. One of his famous exploits was to print up millions of dollars' worth of phoney Erie Railroad certificates and dump them on the market. In the words of Alexander Dana Noyes, "Few properties on which this man laid his hand escaped ruin in the end."

The junk bond king of the nineteenth century was Jay (born Jason) Gould (1836–92). Together with James Fisk and Daniel Drew, codirectors of the Erie Railroad, Gould fought off the attempted 1867 takeover by Cornelius Vanderbilt. By 1872, Gould was deposed from the Erie. He then turned to the West, where he assembled a railroad empire that included the Missouri Pacific and the Kansas Pacific. At one time, he controlled about half the track mileage in the Southwest. Jay Gould was the prototype of the legendary "robber baron."

Government Avenue

Government Avenue, from Broad Street to the Delaware River, is so named because it separates the City of Philadelphia from the federal property of the Philadelphia Navy Yard, built in 1876. Though the city records do not reflect a date for the construction of this roadway, it appears in late nineteenth-century maps.

Governors

The City of Philadelphia has commemorated many of the commonwealth's chief executives by naming streets in their honor. As one travels south, the streets succeed each other in much the same fashion as did their human counterparts. Under the names of individual streets elsewhere in this volume, a reader can usually find more specific information on particular "governor" streets, but a brief summary follows:

Gowen Avenue

Number of Blocks South of Market	Name of Street (Governor)	Term of Office
1900	(Thomas) Mifflin Street	1790–99
2000	(Thomas) McKean Street	1799–1808
2100	(Simon) Snyder Avenue	1808–17
2300	(George) Wolf Street	1829–35
2400	(Joseph) Ritner Street	1835–39
2500	(David R.) Porter Street	1839–45
2600	(Francis R.) Shunk Street	1845–48
2800	(William F.) Johnston Street	1848–52
2900	(William) Bigler Street	1852–55
3000	(James) Pollock Street	1855–58
3100	(William F.) Packer Avenue	1858–61
3200	(Andrew G.) Curtin Street	1861–67
3300	(John W.) Geary Street	1867–73
3400	(John F.) Hartranft Street	1873–79
3600	(Henry M.) Pattison Avenue	1883–87, and 1891–95

THE MISSING GOVERNORS. The only two governors to be skipped by Philadelphia namers were Joseph Heister and John Andrew Schulze. Scuttlebutt has it that neither man provided much in state aid to the city. (Philadelphia Municipal Archives)

There are other Philadelphia streets which bear the names of Pennsylvania governors. However, it appears there was a determined, if unwritten, rule of naming that the streets running south from Mifflin Street would carry the names of the governors in a succeeding order.

Other governors whose names can be found on street signs include Henry M. Hoyt (1879–83), James A. Beaver (1887–91),* and Samuel W. Pennypacker, (1903–7). Five other governors lent their names to streets which have since disappeared. They were Daniel H. Hastings (1895–99), William A. Stone (1899–1903), Edwin S. Stuart (1907–11), John K. Tener (1911–15), and Martin G. Brumbaugh (1915–19). The only two governors to be excluded were Joseph Hiester (1820–23) and John A. Schulze (1823–29). There is no apparent reason for their omission. Ironically, Andrew Gregg, the man who ran against Schulze to replace Hiester, has a street named in his honor.

Gowen Avenue

This Mount Airy street began as Miller Street, between

---

*Beaver can also be included as a name derived from that of a Pennsylvania county. Only the anonymous namer knows for sure.

Germantown and Stenton Avenues, before 1856. James G. Gowen (1790–1873), a native of Ireland, came to America in 1811 and became a merchant. Later, he bought the Miller estate and bred cattle. Sometime after he acquired the property, the roadway became known as Gowen.

Gowen's sons were prominent Philadelphia lawyers. Most well known was Franklin B. Gowen (1836–89), the prosecutor of the Molly Maguires who harrassed and terrorized the miners in the coal regions of Pennsylvania.

Gowen Avenue was confirmed as a roadway in 1878, after the city solicitor opined that the street had been in use for more than twenty-one years. It was not until 1927 that the avenue grew to Rural Lane. It also extended as far as Cheltenham Avenue by 1947.

Gowen Road

See St. George's Road.

Grace Street

See Appletree Street.

Grange Street

Grange Street remembers the Grange Farm, which was once situated near the present-day site of SEPTA's Fern Rock terminal. The first section of Grange Street was deeded to the city in 1891 on both sides of 3rd Street. The street grew between 1909 and 1948 to extend from Wister to Front Street.

Grant Avenue

It has long been held that Grant Avenue was named for Ulysses S. Grant (1822–85), eighteenth president of the

United States. More likely, the name came from Samuel Grant, who purchased land in the vicinity of the present avenue from the Macalesters (see Torresdale Avenue).

The earliest roadbed for Grant Avenue was the old Shady Lane—Pine Road to present-day Shady Lane—opened by jury in 1736. Elements of the old Welsh Road, confirmed in 1823, can also be found in the avenue, from Alton Street and Welsh Road to near Jamison Avenue.

Grant Avenue, by that name, first emerged in 1872, from Tulip Street to State Road. The jury in that year confirmed it from the Wissahickon to the Poquessing creeks. Two years later, it was confirmed from Bustleton Avenue to Blue Grass Road. The remaining sections were included as public roadways between 1896 and 1949.

Gratz Street

Gratz Street, named for the Gratz family, first appeared in city directories as early as 1874, from Oxford to Columbia Avenue.

In view of this dating, it is possible the name came from Simon Gratz (1840–1925). Having graduated from the University of Pennsylvania at the age of fifteen, Gratz was, by the time the street was named, a member of the Pennsylvania House of Representatives (before he reached the age of twenty-one), an assistant city solicitor, and a member of the Philadelphia Board of Public Education. He served on the board of education from 1869 to 1921.

His grandfather, Michael, was a prominent colonial merchant engaged in the India trade. An aunt, Rebecca, is said to have been the model for Sir Walter Scott's character of the same name in *Ivanhoe*. Edward Gratz, his father, was active in the construction of the Pennsylvania Railroad.

By the time of Simon's death, Gratz Street extended its full distance, including old Centennial Street from Oxford to Jefferson Streets, which had its name changed to Gratz in 1895.

GRATZ STREET. Though named for the Gratz family, probably for Simon Gratz, this is an illustration of Simon's grandaunt, Rebecca Gratz, the supposed model for Sir Walter Scott's character of the same name in *Ivanhoe*.

Gravers Lane

John Graver, who owned land in Somerhausen (now Chestnut Hill) after 1800, is commemorated by Gravers Lane. Known earlier as John Streeper's Lane, the roadway—though no records exist—probably changed its name at the time Graver bought the land and opened his popular tavern, later known as the Chestnut Hill Inn and Gold's Hotel.

Opened by affidavit in 1883 from Germantown Avenue to Seminole, the lane first appeared in use by 1864. Union Avenue (Germantown to Stenton Avenue) was added by a name change in 1898; Cathedral (Towanda to St. Martins Lane), in 1959. By 1960, the lane was complete to Cherokee Street.

Grays Ferry Avenue

The Pennsylvania Provincial Council, on October 29, 1696, heard a request for a road "from the lowermost ferry up the Schuylkill commonlie called Benjamin Chambers ferry into the town of Philadelphia." The surveyors were ordered that date to "lay out the Kings road from the said ferry . . . to come into the southermost Street of the town of Philadelphia, and which street running from Delaware River to the Schuylkill." Chambers's, or the Lower, Ferry on Schuylkill was established in the late 1690s, perhaps several months before the petition was filed for the road.

By August 18, 1747, the ferry had changed hands. At that time, George Gray petitioned the council regarding the "road from South street, Grays Ferry to Cobbs Creek, . . . shewing that said road . . . had time out of mind been the only road and accustomed Road to Darby, Chester, Newcastle and the Lower Counties." The petitioners were afraid that improvements to that road had not been properly recorded and asked for a survey and recording. The council's comments were interesting in that they state, "The Road mentioned . . . is an ancient Road in use before the grants of the Province."

When the board resumed consideration on September 8, 1747, the secretary had examined "the Council

Grays Ferry Avenue

GRAYS FERRY AVENUE. Before a bridge spanned the Schuylkill River, travelers had to cross in ferry boats. In 1796, the ferry boat was replaced by a "floating bridge."

books and found therein several orders . . . for the laying out of the several parts of the said road . . . and gave it as his opinion tho' there were no returns of the other parts of the said road to be found on Record yet the whole road had been laid out by order of Council and that it might reasonably be presumed the returns thereof had been given to the late Secretary Patrick Robinson, and that he had omitted to enter them." It seems that after his death many of his records were either lost or destroyed by his widow. The road was resurveyed and entered into the record book.

Gray's Ferry was a very popular spot—so popular, in fact, that a garden was opened in 1790 "fitted out on the plan of the public gardens of London." Contemporaries called it "romantic and delightful beyond the power of description . . . [with] every kind of flower one could think that nature had ever produced and with the utmost display of fancy as well as variety." Even General Washington visited there and dined at the inn. In 1789, the General crossed the ferry on his way from Mount Vernon to New York to be sworn in as president.

By 1796, the ferry boat was replaced by a "floating bridge . . . constructed of logs of wood placed by the side of each other . . . and planks nailed across them." Thomas Twining commented that "although the bridge floated when not charged . . . the weight of our waggons depressed it several inches below the surface. The horses splashing through the waters so that a foot passenger passing at the same time would have been exposed to serious inconvenience."

At the close of the eighteenth century, under the management of Gray's descendents and others (George Weed, George Ogden, Curtis Grubb, and the Kochesbergers), the Garden ceased being the popular amusement spot it once was. It did, however, continue to serve refreshments to travelers.

The "floating bridge" was carried away by the flood of 1789, but it was quickly rebuilt. In 1806, a movement was started to build a more permanent bridge. Gray's "bridge" remained until 1838, when the Philadelphia, Wilmington & Baltimore Railroad, later to become a part of the Pennsylvania Railroad system, built a combination freight and foot traffic bridge on the site. The existing bridge replaced one opened in 1897. The current bridge was completed in 1976.

Gray's Lane

See Cobbs Creek Parkway.

Green's Court

See Quarry Street.

Greene Street

See Counties.

Gregg Street

See Governors.

Gross Street

See Wynnewood Road.

Grubbtown Road

See Crescentville Road.

Guilford Street

See American Street.

Haddington Lane

See Haddington Street.

Haddington Street

The village of Haddington, which formed the western boundary of Blockley Township, became a part of Philadelphia in 1854. Haddington was founded around the Indian Run Creek, a stream which provided the power for woolen mills which sprouted up in and around the village. The street commemorating the village was deeded, from 66th to 67th Street, in 1904. Within twenty years, Haddington Street extended to its present length—from 56th to 67th Street. In 1977, Philadelphia's City Council voted to change the name to Haddington Lane.

Hagys Mill Road

Hagys Mill Road, from Port Royal Avenue to Spring Lane, began as a roadway to Jacob Hagy's grist mill in 1772. Hagy (also spelled Hagge) had petitioned the Quarter Sessions Court for a public road the previous year. As laid out by the surveyors, the road extended from the Ridge Road northwest out of Philadelphia County.

An assessment list of Whitemarsh Township, dated 1780, indicates that Jacob Hagge, William Kagge, Henry Katz, and Henry Sheetz operated paper mills; but the Roxborough history insists Hagge's was a grist mill.

Haines Street

Haines Street was the first Germantown street opened as a street rather than a road. Named for the Haines family, who owned much land along the street as late as 1938, it first appears in records of the Quarter Sessions Court of September 1761. Most of the additions which have extended the street to its present distance were made in the early twentieth century. The name is also apparently a "modern" one, since sections of the street have been known as Bockius (or Pickius) Lane, Methodist Lane, Lafayette, and Meeting House Lane.

The most colorful member of the Haines family was Reuben, who came to Pennsylvania with William Penn. He married Margaret Wister (see Wister Street) and obtained Wyck. The house, which is still standing at 6026 Germantown Avenue, was converted into a hospital for British soldiers after the 1777 Battle of Germantown. Reuben's son Caspar was influential in helping to establish the Germantown and Perkiomen Turnpike Company in 1801 (see Germantown Avenue). On July 20, 1825, while Lafayette was visiting America, Caspar held a reception at Wyck in his honor.

Haldeman Avenue

The old Byberry and Bensalem Turnpike (see Toll Roads), dating back to the late seventeenth century, was renamed Haldeman Avenue in 1907 in honor of a former planning engineer in the streets department.

Extending from Bustleton Avenue to the Roosevelt Boulevard, Haldeman follows the line of the old turnpike as it was confirmed in 1689 and 1848. Sections of the avenue were widened between 1926 and 1962.

Ben Haldeman, who resigned his city post in 1918, died in 1955 at the age of eighty-nine.

Hall Street

See Appletree Street, Beach Street.

Hamilton Street

See Chestnut Street.

Harmony Street

Founded August 24, 1783, the Harmony Fire Company, one of Philadelphia's earliest volunteer fire companies, had its headquarters in a small alley between 3rd and 4th Streets. Since it opened on only one side, the alley was called Harmony Court. In 1895, in an attempt to unify street names, Harmony was changed to Moravian Street. Two years later, history prevailed and the fire company's name returned. (See Fulton Street.)

Harrison Avenue

See Godfrey Avenue.

Harrison Street

Named for William Henry Harrison (1773–1841), ninth president of the United States, according to most folk historians. However, another likely candidate is Thomas Alexander Harrison (1853–1930) a native-born Philadelphian internationally known for his marine and figure painting.

Harrowgate Lane

See Kensington Avenue.

Harry Hosier Way

Harry Hosier was the earliest known black Methodist preacher in America. His first recorded sermon, "The Barren Fig Tree," was preached on May 13, 1781, at Adam's Chapel, Falls Church, Virginia.

"Black Harry," as Hosier was known, was the patron preacher of the African Zoar Church, which was founded at Campington, Philadelphia, in 1794. This is the oldest black congregation within the United Methodist church. Hosier died in May, 1806, and is buried in the Palmer Burying Ground, in the Kensington section.

In 1983, Philadelphia's City Council renamed a part of 12th Street, between Poplar and Market streets, in his honor. As with Richard Allen Way and several other honorifics, Philadelphians still refer to the street as 12th.

Hartranft Street

See Governors.

Haverford Avenue

The early settlers of the Philadelphia area were God-fearing people whose lives centered around their church, or meeting house. Besides, at meeting, or worship, they could meet with friends and neighbors—for perhaps the only social occasion during the week. Thus, reaching the meeting house in Haverford Township, Delaware County, was very important to the Quakers.

Unfortunately, the approaches to John Powell's ferry "were from the south, by a miserable road along the river, and by as poor a road running to the northwest over the hills in a zigzag manner." This "miserable" condition was eliminated by June of 1700, when a new road was opened from Powell's house, passing the Haverford Meeting House, on the way to Goshen, in Chester County. This was the start of today's Haverford Avenue. Originally called the Haverford Road, later it became Haverford Street. It was not until the early twentieth century that the street became an avenue.

Haverford Township was settled in 1682 by Welsh members of the Society of Friends who came from Haverford-West in Pembrokeshire.

Hazelwood Street

See Bonsall Street.

Hazzard Street

Hazzard Street presumably honors the Hazzard family, prominent in Philadelphia affairs since the eighteenth century.

The sections of this street from Kensington Avenue to Coral Street and from Collins to Trenton Avenue were in use before 1883, probably before 1862. In 1895, Hazzard grew to include Fox, from Sydenham to 15th, 13th to 12th, Trenton to Memphis, and Memphis to Gaul (vacated in 1927); and Henderson, from American to Phillips.

The first member of the Hazzard family to achieve prominence was Ebenezer, who became the first secretary of the Insurance Company of North America in 1795. His sons were well known in their own right—Samuel, as an authority on Philadelphia history, and Erskine, with the Lehigh Navigation Company.

Henry Street

See Keyser Street.

Hepburn Street

See Gorgas Lane.

Hermit Street

Johannes Kelpius was an educated, affluent refugee who came to America from Germany in 1694 to await the Judgment Day. He and his followers settled in Germantown. But finding that they and their religious thinking attracted too much attention, they moved to the banks of the Wissahickon Creek and lived in caves. Here the "Hermits of the Mystic Brotherhood," more popularly called the "Hermits of the Ridge," practiced and taught magic, divining, mathematics, astronomy, and science to anyone who would listen.

Kelpius' monks held open-air concerts to lure the

"Women of the Wilderness" out of the woods and into their arms. Such a woman is mentioned in the Book of Revelations "as she was to come from the wilderness, leaning on her beloved." After years of waiting—and celibacy—some of the weary monks substituted the more substantial ladies of Germantown for their Ideal Woman. This resulted in great conflict and disagreement within the religious order. The monks, having tasted the joys of the flesh, drifted back into the normal humdrum of everyday life.

But Judgment Day and other tenets of their faith did not come about as predicted. Kelpius, who thought he would not die but be carried bodily to heaven, expired in 1708 at the age of thirty-five. His group disintegrated after his death. Some of his followers returned to Germantown; others went to Ephrata, Pennsylvania. But Kelpius's cave, though well hidden, still exists, and is marked with a granite engraving. It is located a few hundred yards inside Fairmount Park, west of Hermit Terrace.

The earliest section of present-day Hermit Street was confirmed in 1804, from Ridge Avenue to the Park at Rittenhouse Street. The span from Cresson Street to Ridge was deeded to the city in 1871.

Hermit Terrace

See Hermit Street.

Hermitage Street

See Galloway Street.

Heston Street

Named for a section of West Philadelphia situated about one mile south of George's Hill in Fairmount Park, Heston Street, from Paxon to Wilton, was partially deeded to the city as early as 1895. A small section, 290 feet west from 52nd Street, was apparently in use before 1883.

Heston Street

Hestonville was named for the Heston family, who settled that area about 1800.

Hibberd Street

See Sartain Street.

Hickory Lane

See Fairmount Avenue.

Hicks Street

This street was named for Elias Hicks (1748–1830), a leader of the Society of Friends (the Quakers).

Hicks was an early mover in the drive to abolish slavery. In 1811, he published *Observations on the Slavery of the Africans and Their Descendents*, in which he called for a boycott of all products related in any way to slavery. He was also a spokesman for the quietist and individualist side of Quakerism that saw salvation as the result of each person following one's own "inner light" through which God's ongoing revelation is achieved. He opposed the evangelism that was sweeping the Quakers. At the 1817 yearly meeting in Baltimore, he successfully resisted the adoption of a creed. Hicks' influence led to the 1827–28 schism. The liberal branch became known as the Hicksites, even though Elias Hicks did not take part in the split.

High Street

See Market Street.

Hoffman Street

See Noble Street.

Hog Island Road

This road got its name because it led to Hog Island. The island is now part of the mainland—and the road has almost disappeared.

As early as 1750, the island in the Delaware was called Hog. It is possible that this was the same island that an early Swedish cartographer called Keyser in 1654. The Indians, long before the Swedes and the English, called the island Quistconk. The Lenape Indian word for hog is *goschgosch* and the Swedish word is *kwskus*. This might indicate the Indians, in a gesture of friendship, tried to give the island a Swedish name.

On the other hand, the Indians' naming might have indicated something entirely different. And the third Swedish minister to the colony on the Delaware was Israel Holg; his name might have been mispronounced by the English.

The pinnacle of Hog Island's fame came during World War I, when a mammoth shipyard was erected there in 1917–18, almost overnight. More important is the folk tradition that the Philadelphia hoagie got its start there. It seems the "antipasto sandwiches" carried by the workers to the shipyard took on the name of "hogies"—after their destination. Mispronunciation and misspelling brought the word around to present-day "hoagies."

Holme Avenue

See Holmesburg Avenue.

Holme's Street

See Arch Street.

Holmesburg Avenue

On the 1681 map of Penn's province, a small section is

marked as being the property of Elenor and Thomas Holme. That property of Penn's surveyor-general became a village in Lower Dublin Township. Holmesburg was made a part of the City of Philadelphia in 1854.

Originally called Race Road, the avenue became Holmesburg in 1895, from Frankford Avenue to Torresdale. No record exists to indicate how long Race had been in use before that year. Other sections of Holmesburg Avenue have existed physically through the years—from Solly Avenue to Frankford and from Hegerman to the Delaware River—but they have since disappeared from the city plan.

Holme Avenue was also named in honor of the surveyor.

House Numbering

For more than a century after William Penn landed, Philadelphians did not have numbers on their houses. That was fine when Penn's "Green Country Towne" was just that. The trouble began when Philadelphia started to emerge as a major city.

Captain John Macpherson (Macpherson Street bears his name), a methodical madman, tried to get the city fathers interested in placing numbers on houses . . . but without success. The captain wanted the legislation passed because he was publishing a street directory, and it would be of little use without numbers to differentiate one house from another.

Macpherson's directory, the first street directory in the United States, was published in 1785. Unbeknownst to the lawmakers, Macpherson had adopted his own peculiar system of numbering: He started out on the south side of one street and numbered the houses 1, 2, 3, and so forth. When he reached the end of the built-up portion of the city, about 5th or 6th Streets, he started back on the north side. In other words, houses with the lowest and highest numbers on any street would face each other.

Macpherson's system didn't last long. In 1790, Clement Biddle, the United States Marshall, conducted the first census. In the course of his assignment, he renumbered the houses with even numbers on the south and west; odd numbers on the north and east. If a vacant lot existed, Biddle skipped it and numbered the next house.

Biddle's numbering system later produced house numbers with halfs and quarters. The marshall used the information he gained to publish another street directory. The city used Biddle's system until the middle of the nineteenth century, when Councilman John F. Mascher introduced a radical new system: the centurial method. Mascher Street, by the way, bears his name.

Under Mascher's format, each city block was assigned one hundred numbers, which would be divided by the average building lot size. He promoted his idea by saying, "A navigator calculates the position of his ship by observations made on the compass and the position of the stars. So also a pedestrian may at any moment know at what point in the city he may be, by merely looking at those constantly recurring landmarks [house numbers], only with this advantage, that it requires no calculation on his part whatsoever."

Mascher was right, and Philadelphia had the first intelligent approach to house numbering in the nation. Mascher's system, the one that is used today, is easy to understand but, like many other things Philadelphian, it took a long time coming.

Howard Street

See Ruffner Street.

Howell Street

See Cherry Street, Delancey Street.

Hoyt Terrace

See Governors.

Hudson's Lane

See Christian Street.

Hunting Park Avenue

Hunting Park Avenue was named for the park through which it runs. Once the great hunting grounds of the Lenni-Lenape Indians, Hunting Park was used as a race track, opened in 1808 and known as Allen's Race Course. The name was later changed to the Hunting Park Race Course. Because of laws prohibiting horse racing, the course was gradually abandoned. In 1854, a number of public-spirited citizens purchased the land and donated it to the city for "the use of the public as a park." Two years later, the ground was dedicated "free of access for all the inhabitants of the city, and for the health and enjoyment of the people forever under the name of Hunting Park."

In 1871, the Fairmount Park commissioners were authorized to open a street between Fairmount Park and the Hunting Park. There was already a road, known as Nicetown Lane, which had been in use from the Wissahickon to the Germantown Road since 1742. This roadway was further confirmed in 1820 from Germantown to the Schuylkill River. Nicetown Lane was extended as far as Frankford Avenue in 1784. The name was changed to Hunting Park Avenue by the authorizing legislation of 1871.

Hunting Park Avenue, though confirmed earlier, was completed to its present extent—East River Drive to Frankford Avenue—by 1923. Though officially known as Hunting Park Avenue, the first hundred yards of roadway off the drive are still marked by the streets department as Nicetown Lane.

Huntingdon Street

See Counties, Race Street.

Huron Street

See Indian Tribes.

Illinois Street

See Bouvier Street.

Independence Square

See Squares.

INDEPENDENCE MALL. In 1950, it was difficult to locate Independence Hall amid the clutter of buildings and factories. As part of Philadelphia's urban renaissance, extraneous buildings were demolished and a beautiful tree-lined mall was created. (Philadelphia City Planning Commission)

Indian Queen Lane

Indian Queen Lane commemorates either the signboard of an early tavern along its route or an Indian legend. Finding no record of a tavern in the vicinity—the closest being the Indian Queen at Germantown Avenue and Queen Lane—we must assume the name commemorates a legend.

The road was first confirmed in 1773, from Germantown to Ridge. Indian Queen Lane remained virtually unchanged until 1918, when it was widened from Vaux to Conrad Street. The section from Henry Avenue to Vaux was dedicated in 1925. The earlier sections were widened, from Vaux to Cresson, between 1947 and 1953. (See Queen Lane.)

Indian Tribes

Perhaps because William Penn had a great reputation as a friend of the American Indian, a number of Philadelphia streets bear the names of tribes. There is no record of Penn's ever suggesting or assigning a tribal name to a street. In fact, the use of these names seems to be of recent origin.

The most intensive tribal name adoption took place in 1900 in the Chestnut Hill area, when the names of several numbered streets were altered: 27th Street became Shawnee Street, 29th became Navahoe (now Navajo) Street, 30th became Seminole Avenue, 33rd became Huron Street, and 35th Street became Cherokee Street. There is really no reason for these particular name changes. The tribes listed have neither local significance nor geographic similarities. Probably they were named in that manner simply because they sounded good.

Other street names derived from tribes are Cayuga Street, Chippewa Road, Comanche Road, Dakota Street, Erie Avenue, Iroquois Lane, Mohican Street, Narragansett Street, and Osage Avenue. Of all the tribes commemorated, only the Eries, Iroquois, and Shawnees have the slightest connection with Philadelphia or Pennsylvania history: the Eries and Shawnees were located in Pennsylvania before the famous "Walking Purchase" of 1737; the Iroquois were the main force allowing the Pennsylvania tribes to live in peace with Penn. But though Erie is a name for a tribe, the avenue most probably was not named for the Indians but for the Pennsylvania county.

Indiana Avenue

Named for the Pennsylvania county (see Counties), Indiana Avenue was confirmed by the road jury from Kensington Avenue to Front Street in 1872. To the east, the avenue was completed by the inclusion of Neff Street, from Allen Street to Kensington Avenue, in 1906, and to the west, as far as 35th Street, by 1929.

Indiana County, formed in 1803, supposedly was named for the Territory of Indiana, after "one of the old ante-Revolutionary land companies," the Indiana Land Company.

Ingersoll Street

This street honors a family which has served the City of Philadelphia from its beginning up to, and including, the present. The particular member usually cited for the honor is Jared Ingersoll (1749–1822), a Philadelphia lawyer and signer of the Constitution.

The oldest recorded section of contemporary Ingersoll Street is the block from 24th to 23rd Street, deeded by William Bucknell (see Bucknell Street) in 1854, but since vacated. In 1895, Ingersoll acquired Seybert (Ridge Avenue to 18th Street), Stein (18th to 17th), and Eliza (Smedley to 15th). The remainder, as far west as Taney Street, was deeded by 1922.

A much earlier section of this street, from Front Street to Frankford, not found in streets department records, was called Jackson Avenue until 1858.

Iroquois Lane

See Indian Tribes.

Irving Street

See Rittenhouse Square.

Iseminger Street

Honoring a South Philadelphia family active in the building trades in the mid-nineteenth century, Iseminger Street first appeared under that name in 1858, between Spruce and Pine, east of 13th Street, when Jackson Street underwent a name change.

Among the first Isemingers was Adam, a wheelwright who lived at Queen Street, below 6th, in 1800.

Iseminger Street

The best known, however, was Schubert Iseminger, a housepainter, in 1850.

Before the end of the nineteenth century, Iseminger Street had grown to include Dean Street, from Shunk to Wharton; Stockton, from Lombard to Waverly; and Pike, from Cherry to Quarry. Its present length, from Bigler to Berks, was achieved by 1923.

Jackson Street

This honorific is generally assigned to Andrew Jackson (1767–1845), seventh president of the United States. But Abraham Reeves Jackson (1827–92) was a Philadelphia physician and pioneer gynecologist who was immortalized by Mark Twain as the doctor in *Innocents Abroad*. (See Iseminger Street.)

Jacoby Street

See Camac Street.

James Street

See Chestnut Street, Noble Street.

Jefferson Avenue

See Moyamensing Avenue.

Jefferson Street

Jefferson Street honors the drafter of the Declaration of Independence and third president of the United States—second hand. Named after the Pennsylvania county (see Counties) established in 1804 out of Lycoming County, Jefferson Street was dedicated, from Ridge Avenue to 24th, in 1850. Four years later, the road jury confirmed it as far as 33rd Street. The street grew first to the east—to 6th Street by 1850—and finally, by acquisition of Sage Street, as far as Frankford Avenue. To the west, progress was slower. It was not until 1911 that Jefferson reached to Atwood Road.

JEFFERSON STREET. Drafter of the Declaration of Independence and third President of the United States, Thomas Jefferson proved the mythic quality of Holme's map of Philadelphia. When asked about the house in which he wrote the Declaration, Jefferson described it as being the last house in town before the wooded areas. The house was on 7th Street! (Bureau of Engravings)

John F. Kennedy Boulevard

America was shaken to its foundations by the assassination of President John F. Kennedy on November 22, 1963. As civic leaders throughout the nation searched for ways to show their sadness, it became commonplace in almost every city, town, or hamlet to rededicate a major thoroughfare in Kennedy's honor. Philadelphia was no exception.

On December 18, 1963, Mayor James H. J. Tate, a political ally of Kennedy's, signed into law an ordinance "to perpetuate the memory of this courageous man who visited this City on many occasions" and who "had endeared himself to the hearts of the American people, and especially to the hearts of the citizens of the City of Philadelphia." The ordinance authorized that Pennsylvania Boulevard, from Juniper Street to Schuylkill Avenue west of the Pennsylvania Railroad station, then from 30th Street to the intersection of 32nd and Market Streets, be changed to "John F. Kennedy Boulevard."

Less than three years later, the newly constructed park, in the shadow of the City's Municipal Services Building, bounded by Arch, 15th, Kennedy Boulevard, and 16th Street, was renamed John F. Kennedy Plaza.

John Gloucester Way

See Girard Avenue.

John Steeper's Lane

See Gravers Lane.

Johnson Street

Though folk historians take the Johnson of this street to be the seventeenth president of the United States, Andrew Johnson (1808–75), it is more likely that the name was derived from one of the original settlers of Germantown.

Dirck Jansen came to this country from Holland and settled in Germantown. In 1765, he began construction of a house for his son John Johnson. Completed three years later, the house apparently impressed John, who brought his bride, Rachel Livezey (see Livezey Street), directly to it after exchanging vows at Meeting. John died in 1805 and his son Samuel inherited the house. During the Civil War, the house is reputed to have been a station on the Underground Railroad.

The earliest section of Johnson existed from Musgrave to Wayne Avenue before the Civil War. From Wissahickon to Chew Avenue, the street was opened by affidavit between 1885 and 1901. In 1895, old Nice Street, from Walnut Lane to Wissahickon Avenue, became Johnson. The remaining blocks, from Chew to 75th Avenue North, were included by 1924.

Johnston Street

See Governors.

Jones Alley

See Church Street.

Joseph Kelly Terrace

See Pelle Road.

Juniata Street

Named for the Pennsylvania county of the same name (see Counties), Juniata Street was first deeded to the city, from Clarissa to 16th Street, in 1877. Interestingly enough, though the street has included other sections—it once extended from Henry Avenue to the Delaware

Juniata Street

River with several interruptions—the original section is all that remains.

Juniata County, formed in 1831, took its name from the Juniata River. *Juniata* is an Iroquois word which means "standing stone people" or "the people of the standing rock."

Kansas Street

See Catharine Street.

Kelly Drive

On March 14, 1985, the Fairmount Park Commission renamed the 4.5-mile East River Drive, that scenic roadway that starts at the Art Museum and winds along the Schuylkill's east shore to Lincoln Drive. The name selected was Kelly Drive.

Kelly Drive honors two men who were intimately involved with Philadelphia's Boathouse Row and Fairmount Park. The Kellys were Olympic medalists who spent countless hours on the river.

John B. Kelly, Jr. (Kel to his friends) died in 1985 of a heart attack while jogging in Center City. He and members of the Fairmount Rowing Association had just completed a six-mile row on the river. Kel, a former city councilman, had recently become president of the U. S. Olympic Committee.

The senior Kelly had been a Fairmount Park commissioner for many years, and was the father of Princess Grace of Monaco. When Grace became engaged to Prince Ranier, Kelly noted that his kingdom—Fairmount Park—was larger than his intended son-in-law's. The swimming pool next to Memorial Hall is named in the senior Kelly's honor.

Changing the name of such a popular roadway does have its problems. Philadelphia's chief traffic engineer, Jack Boorse, anticipated that some diehard Philadelphians would continue to call the drive by its old name. "The Avenue of the Americas in Manhattan," he said, "has been around for years, and people still call it Sixth Avenue." But diehards weren't the only problem . . . (See Palumbo Square.)

Kelton Street

See Carlisle Street.

Kennedy Boulevard

See John F. Kennedy Boulevard.

Kensington Avenue

The principal street of the town of Kensington adopted the community's name for its own. Kensington Avenue first appeared in the street directory of 1851. However, it must have existed much earlier as part of the road to Frankford. Early in the nineteenth century, it was the road to Harrowgate Springs, a popular spa of the period, from which springs the name of Harrowgate Lane.

Officially, the first section of Kensington was confirmed by jury in 1847 from Front Street through the Norris estate to the Leamy estate. Added to mostly by use rather than by deed or ordinance, Kensington Avenue reached its full distance before 1890. In 1857, the avenue was a turnpike (see Toll Roads). Between 1860 and 1870, it was known as the Frankford Road.

The town of Kensington was laid out by Anthony Palmer (c. 1675–1749) and named for the section of London. Because of the principal industry of the residents, Kensington was nicknamed Fishtown—a name which continues in use to this day. Palmer, by the way, has a Kensington street named in his honor.

Kershaw Street

See Appletree Street.

Keyser Street

Keyser Street, from Wyneva to Price Street, honors the Keyser family.

Dirck Keyser came to America from Amsterdam as early as 1688. He was one of the early Mennonite founders of Germantown. Together with his son, Keyser built a two-story house, which is still standing, at 6205 Germantown Avenue.

Keyser Street came into existence in 1895, when Henry Street, from Wyneva to Hansberry, underwent a name change. Two years later, the street grew to include old Drinkwater Street.

King Street

See Water Street.

Kingsessing Avenue

Kingsessing Avenue takes its name from Kingsessing (now part of southwest Philadelphia), one of the oldest settled regions of Philadelphia. The name Kingsessing is a corruption of the Indian *chingsessing*, "a place where there is a meadow."

The earliest recording of Kingsessing Avenue appears to be a deed of dedication, dated 1858, for a section west of 67th Street, vacated in 1956. The line of the avenue was confirmed by a resolution of council in 1869 from 46th Street to 71st. Apparently, construction did not meet expectations because two years later council legislated the extension of the road from 63rd to 71st Street. The remainder of Kingsessing Avenue, from 42nd to 45th Street, was confirmed by jury in 1894. That section was vacated between 1968 and 1970.

Kingsessing was the "place on the Schuylkill where five families of freedom dwelt together" and is considered by some authorities to be the first village of Philadelphia.

Kirkbride Street

See Bridge Street.

Knight's (Knecht's) Court

See Appletree Street.

Knights Road

Named for Giles Knight, who came over on the *Welcome* with William Penn.

Knorr Street

See Potter Street.

Knox Street

Honors Henry Knox (1750–1806), a general during the Revolution.

Ladley Street

See Rittenhouse Street.

Lafayette Lane

See Haines Street.

Lambdin Street

See Bouvier Street.

Lancaster Avenue

Very few years had passed from the time Penn established his city on the river before hardy pioneers pushed westward into the fertile farmland of Lancaster County.

In 1730/31, the residents of that area petitioned the "Governour in Council" for a road so they might "bring the produce of their labor to Philadelphia." By June 1733, the proposed roadway was viewed and the order was given that the "King's Highway or Publick Road . . . be forthwith cleared, and rendered commodious for the Public Service."

The road began shortly thereafter. Axemen began clearing the way through the woods. Streams were crossed by the erection of rough-hewn log bridges or fords. But it was not until 1741 that the road was opened—for foot traffic.

In 1792, the legislature authorized a company, formed the year before, to construct a turnpike from Market Street in Philadelphia to the City of Lancaster. Though the roadway was supposedly completed two years later, the first regular stage, carrying ten passengers, did not use the road until May 1797. The journey of sixty-five miles took a total of twelve hours. It was "a masterpiece of its kind," Francis Bailed declared in his journal in 1796. Indeed, Old Lancaster Road was the first turnpike, and the first paved highway in America.

Lancaster Avenue

Before the road to Lancaster was freed from toll (see Toll Roads) in 1903, the City of Philadelphia had acquired land along the avenue's present route within the city limits. By 1936, Lancaster Avenue, as known today —from 35th Street to City Avenue—was complete.

The starting point of the old road at 32nd Street and several other sections as far west as 35th Street were vacated to accommodate the twentieth-century expansion of Drexel University.

Lancaster Street

See Norris Street.

Landreth Street

See Quarry Street.

Lansdowne Avenue

John Penn's country seat, Lansdowne, was situated on the west bank of the Schuylkill River. It was on what was the original Lansdowne estate that the principal buildings of the 1876 Centennial Exposition were built. The present-day avenue begins approximately at the western boundary of the old Penn grounds.

The grandson of Philadelphia's founder, John Penn completed his mansion in 1777. He gave the estate the name of Lansdowne to honor William Petty, the first Earl of Shelburne, who later became the Marquis of Lansdowne. To be sure, Petty did not receive this title until seven years after the estate was named. As Earl of Shelburne, however, Petty did live in Lansdowne House in London.

William Bingham acquired the property for his summer home in 1797, after Penn had returned to England. In 1804, it was purchased by the Baring family (see Baring Street). Joseph Bonaparte, ex-King of Spain, leased it in 1816 and lived there for two years.

Lawson Street

The mansion was accidentally destroyed by fire in 1854 by several young children, including a future pioneer of the city of Denver, who were experimenting with fireworks for the Fourth of July. The estate lands and the walls of the old mansion were conveyed to the City of Philadelphia in 1866 and became part of Fairmount Park.

LANSDOWNE AVENUE. Contemporary Lansdowne Avenue begins at the western boundary of John Penn's country estate. Named in honor of William Petty, the first Earl of Shelburne, who lived in Lansdowne House in London, Penn's summer house was completed in 1777. It was later lived in by William Bingham, the Baring family and Joseph Bonaparte. The house was destroyed by fire in 1854.

Laurens Street

Laurens Street honors the Laurens family. The most notable members were Henry (1724–92), a Revolutionary War statesman, and John (1754–82), an officer under Washington.

Lawrence Street

See Counties.

Lawson Street

See Moravian Street.

LAURENS STREET. Henry Laurens succeeded John Hancock as president of the Continental Congress. He resigned following the Silas Deane affair, and was selected to negotiate a treaty of "commerce and friendship" with Holland. He was captured by the British and held prisoner for 15 months. Released on bail, he was finally exchanged for Cornwallis in 1782.
(Painting by John S. Copley, National Archives)

League Street

From League Street to Government Avenue, the start of the property of the Philadelphia Naval Base, is almost three miles—a league away. About the time the first block of League was dedicated to public use in 1863, anticipation was high that the city would be favored by the federal government building the newest, most modern shipyard in the nation. The new Philadelphia Naval Shipyard, replacing an earlier one at the Delaware foot of Federal Street, was finally built in 1876.

By the end of the nineteenth century, League Street had increased in size, though it never achieved a league in length, by the change of the names of the following streets: Albert, from 12th to 13th (since vacated); Suffolk, from 9th to Passyunk Avenue; Paul, from 6th to Fairhill; Mechanic, from Reese to Randolph; Clark, from 3rd to 4th (since vacated); Mary, from Front to 2nd; and Reckless, from Water to Front (vacated for the Delaware Expressway).

Lebanon Avenue

See Counties.

Lee Street

The street was named for the Lee family of Virginia, whose members include Charles (1731–82), Revolutionary War general; Francis Lightfoot (1734–97), signer of the Declaration of Independence; Henry "Light Horse Harry" (1756–1818), Revolutionary War officer; Richard Henry (1732–94), signer of the Declaration of Independence; and Robert Edward (1807–70), commanding general of the Confederate forces during the Civil War.

Lehigh Avenue

First dedicated to the city in 1850 from Belgrade to Coral Street, Lehigh Avenue was named for the Pennsylvania county (see Counties).

The commissioners of Kensington opened the next section—from 6th Street to Germantown Avenue—in 1854. By the end of the nineteenth century, Lehigh had reached its full length, from Richmond Street to Ridge Avenue.

Lehigh County was organized in 1812. It took its name from the Lehigh River, which separates Lehigh from Northampton County. The Delaware Indians called the river Lechauwekink, "where there are forks." German settlers shortened the name to Lecha and the English pronounced it Lehigh.

Lehman Street

See Price Street.

Leib Street

See Blair Street.

Leiper Street

Thomas Leiper's plan to cut a canal from his stone quarry to the tidewater was rejected as "chimerical, visionary and ruinous." Undaunted by these comments, Leiper created—instead of a canal—the first practical railroad in America.

Leiper came to this country from Scotland in 1764. By 1774, he was manufacturing snuff at 274 Market Street. Later, he opened a stone quarry on Crum Creek, where he also built a snuff mill and a grist mill. The railroad, operating from 1809 to 1828, ran for one mile from his quarry to Ridley Creek. Leiper's mansion, built in 1785 and called Avondale Place, has also been commemorated in a street name.

The street that bears Thomas Leiper's name was first established from Oxford to Church Street in 1831. Added to in 1891, 1916, and 1922, it now runs from Oxford to Pratt. (See Camac Avenue.)

Letitia Street

This little street, extending from Chestnut to Market, was named in honor of William Penn's daughter. Penn "deeded the block extending on Market Street from Front to Second and halfway to Chestnut Street" to the girl in 1701.

The lot of ground was bought by Thomas Chalkley for investment purposes in 1707. A small house built on the street between 1713 and 1715 supposedly was the home of Penn's daughter. This claim has been disproved. The Letitia Street House was moved to its present location—above Girard Avenue in Fairmount Park—in 1883.

The roadway from Letitia Court to Black Horse Alley was confirmed in 1827; the remainder, in 1855. In 1895, Letitia Street became Mascher. Two years later, an ordinance was passed to restore the original name.

Levering Avenue

See Monastery Avenue.

Levering Street

Levering Street, as originally confirmed in 1851, runs from Main Street to Tower and was named for the Levering family. Wigard and Gerhard Levering came to Germantown in 1685 from Münster, Germany. Five years later, they founded Roxborough. William, grandson of Wigard (or Wigart), owned considerable land in that area and opened the first inn in Roxborough at the "Sign of the Tun," later called the Leverington Hotel. Through his efforts, the first schoolhouse in Roxborough was built. "For and in consideration of the love and regard they have and bear for the public good in having a school kept in their neighborhood," he and his wife granted twenty perches (perch = 5½ yards) of land to seven trustees for the Roxborough School. The William Levering School, at Gerhard and Ridge, continues in his memory.

Leverington Avenue

Leverington Avenue derives its name from the village of Leverington, once situated on Ridge Avenue east of Manayunk between Allens and Gorgas Lanes. The village adopted the name of the Levering family (see Levering Street), the first settlers of Roxborough.

The earliest recording of this avenue, then known as Washington Street, was in 1822, when the jury confirmed the road from Canal to Fountain Street. The next year, the road continued as far as Cinnaminson. By the end of the nineteenth century, Leverington Avenue extended from Flat Rock Road to Lawnton Street. The remainder—to Henry Avenue—was opened by ordinance in 1947.

Lewis Street

See Moravian Street.

Library Street

Library Street, now considered part of the Independence National Historical Park, was confirmed as a public road in 1809. Named for the building erected on it in 1789–90 for the Library Company of Philadelphia, the street runs only from 4th to 5th Street.

The Library Company, formed in 1731 by Benjamin Franklin, is the oldest circulating library in the nation. The collection of the Library Company is now housed on Locust Street west of 13th. The present building of the Library of the American Philosophical Society on Library Street, though a reconstruction, contains an extensive collection of Franklin's important papers.

In 1895, Library's name was changed to Sansom. In 1952, the original name was restored.

Limekiln Turnpike

Taking its name from the lime quarries of Montgomery

Limekiln Turnpike

County, the road's first terminus, Limekiln Turnpike originally was granted for use in 1693 from the lime quarries of Thomas Fitzwater in Upper Dublin to Haines Street in Germantown. In 1735, the jury confirmed Limekiln Road from York Road to the county line. The road, or turnpike, existed as a toll road until 1903 (see Toll Roads). Except for widening operations in the twentieth century, Limekiln Turnpike exists today much as it did in the late seventeenth century.

Lincoln Drive

This roadway was named in honor of Abraham Lincoln (1809–65), sixteenth president of the United States. A statue of Lincoln signing the Emancipation Proclamation sits in the middle of the road across from Boat House Row. The date of the signing on the statue is wrong!

Lindbergh Boulevard

Charles Augustus Lindbergh (1902–74) was a barnstorming pilot until May 20, 1927, when he flew *The Spirit of St. Louis* from Long Island's Roosevelt Field to LeBourget airfield, near Paris. The 33½ hour trip made "Lucky Lindy" a national celebrity. As a result of his acclaim, cities across the United States named streets and roads in his honor.

In 1932, Lindbergh's two-year-old son was kidnapped and later found murdered. Publicity made the kidnapping and the trial of Bruno Richard Hauptmann the news story of the century, and pushed the Lindbergh family into self-imposed exile in Europe. Prior to World War II, Lindbergh, as a member of the isolationist America First Committee, urged America not to enter the European war. For this he was sharply criticized as a Nazi sympathizer.

His account of his Paris flight, *The Spirit of St. Louis*, received the Pulitzer Prize in 1953.

Lindsay Street

See Bancroft Street.

Lingo Street

See Bouvier Street.

Little George Street

See Sansom Street.

Livezey Lane

See Livezey Street.

Livezey Street

In 1746, the jury confirmed a road to the mill of Thomas Shoemaker. This road extended from the present-day line of Fairmount Park to Cresheim Valley Drive. The

LIVEZEY STREET. Thomas Livezey was a miller who helped found the Union School House of Germantown, now Germantown Academy. He was also noted as a producer of good wines, as comments from satisfied customer Benjamin Franklin indicate. Shown is the Livezey House.

Livezey Street

very next year, Shoemaker sold his mill to Thomas Livezey.

Thomas Livezey was a man of many talents. He lived beside his mill and produced, in addition to mill items, good wine. Robert Wharton sent a selection to Benjamin Franklin, who wrote back: "I received your favours . . . with another dozen bottles of excellent wine, the manufacture of our friend Livezey." The miller helped found the Union School House of Germantown (now Germantown Academy) in 1759. He served as a justice of the peace and, in 1765, as a provincial commissioner. His daughter Rachel married John Johnson (see Johnson Street).

Prior to 1958, Livezey Lane was designated as a street.

Lodge Street

See Sansom Street.

Logan Square

See Squares.

Logan Street

See Wingohocking Street.

Lombard Street

The site of Philadelphia's first financial district, Lombard began as an alley in 1740 from Front to 2nd Street. When the street was finally opened by affidavit in 1883 from the Delaware to the Schuylkill River, there had been only one prior legal recording of any opening of Lombard: the section from Front Street to Delaware Avenue was confirmed by the road jury in 1835.

Lombard Street obtained its name from London's

LOGAN STREET. James Logan, William Penn's colonial secretary, donated his name to many things . . . Logan Street, Logan Circle, and others. He did not, however, swap his name with Chief Wingohocking. (Scharf and Wescott's *History of Philadelphia, 1609–1884*)

financial street. The name is originally derived from the Italian moneylenders of Genoa and Florence—the Lombards—who replaced the persecuted Jews in the fourteenth and fifteenth centuries.

Loudon Street

See Armat Street.

Luzerne Street

See Counties.

Lycoming Street

The roadbed of the old Nicetown Lane, confirmed in 1789, travels through Lycoming Street from Germantown Avenue to Frankford. Named for the Pennsylvania county (see Counties), the street became known as Lycoming in 1895, when earlier streets—including Barr, from Germantown Avenue to Broad Street, and Roxborough, from Broad Street to Old York Road—lost their original names. Lycoming was completed to its full extent by 1921, from Germantown Avenue to Hunting Park.

The county, formed in 1795, was named for the Lycoming Creek, which separated the settled part of Northumberland County from the Indian territories. The Indian word *lycomin* means "sandy or gravelly creek."

Macalester Street

Charles A. Macalester, son of a Scottish sea captain and prominent Philadelphian, was an active real estate developer, civic leader, and entrepreneur of the mid-nineteenth century. In 1866, Macalester was at the center of a controversy regarding the right of an owner of ground on both sides of an alley to possess the ground within the line of the alley. His particular case, championed by Eli Kirk Price (see Price Street), concerned itself with the properties at 315–319 South 12th Street. Price won the case—a landmark at the time—and Macalester was able to keep the house he had built at 317.

Macalester Street, from Hunting Park Avenue to Cayuga Street, opened by ordinance in 1924, commemorates Charles A. Macalester.

Macpherson Street

See House Numbering.

MOUNT PLEASANT AVENUE. In 1761, Captain John Macpherson bought land in what is today's Fairmount Park. Using "the spoils of his privateering," he built "Clunie" as his country seat. Later, he changed the name to "Mount Pleasant." In 1779, Macpherson sold the estate to Benedict Arnold, who gave it as a wedding gift to his bride, Peggy Shippen. After Arnold's treason, the house passed through several hands until, in the mid-19th century, Mount Pleasant and the surrounding estates were acquired by the City of Philadelphia and made part of Fairmount Park.

Main Street

See Flat Rock Road.

Manayunk Avenue

Named after the section of Philadelphia called Manayunk, this avenue extends from Rochelle Avenue to Fountain Street.

The section, originally called Flat Rock, grew up between 1818 and 1821 with the construction of a dam, canal, and locks by the Schuylkill Navigation Company. At a meeting of the inhabitants of Flat Rock, held November 3, 1824, it was resolved to designate the village Manayunk. Incorporated as a borough in 1840, it became part of Philadelphia following the 1854 consolidation (see Consolidation).

Parts of Manayunk Avenue were deeded to the city as early as 1871. It was not complete to its present extent until 1926.

The name Manayunk has been defined by authorities as "where we go to drink" from the Delaware Indian *mene-iunk*, another name for the Schuylkill. Another possibility is the Lenape name for the group of islands at the mouth of the Schuylkill—*menateyonk*, "at the islands."

Manderson Street

Manderson Street is a victim of progress. Since first declared legally open by affidavit in 1894, from Frankford Avenue to Beach Street, Manderson Street has gradually and systematically been reduced in size. Though still extending from Frankford to Beach, the street has lost ground because of the expansion of Frankford and Delaware Avenues, and improvements made to both.

The man who gave his name to Manderson was the owner of Petty's Island, located in the middle of the Delaware on a line with his street.

Manheim Street

"One of the English weavers," an account reads, "said that Manheim Street was originally a cart track to the farms in the rear [of Germantown]. A part of it was called Shinbone Alley." Another account avers it was opened by the Shippen family about 1740. On old deeds for the area, it is listed as Shippen's Lane, Pickus's Lane, Betton's Lane, and Cox's Lane.

Manheim Street, as such, did not come into legal existence until 1803, when it was confirmed by the Quarter Sessions Court from the Germantown Road to the Wissahickon Township line—its present distance (Germantown Avenue to Wissahickon Avenue).

There is a question as to where the name was obtained. Two suggestions are most popular. The first is that Jacques Roset named it for the ladies of Manheim, Germany. Unfortunately for those romantics who subscribe to this theory, the street's name appears as Manheim on a 1780 deed—twelve years before Roset's arrival. The second attributes it to Henry Fraley, who laid out the town of Manheim in Germantown in 1796. Again, the name appears before the act.

It is more likely that the name honors the home of Baron Heinrich Wilhelm von Stiegel, who upon his arrival in America from Manheim, Germany, in 1761, purchased considerable land in Germantown. He also bought two hundred acres of land in Lancaster County, erected his glass works, and called the town Manheim.

Manning Street

See Bach Place.

Manship Street

See Sartain Street.

Mantua Avenue

The village of Mantua, north of Spring Garden Street and northeast of Lancaster Avenue, took its name from Mantua, Italy, the home, supposedly, of Virgil. Judge Peters of Belmont (see Belmont Avenue) laid out the village in 1809. It was not, however, until long after Mantua Village became a part of the City of Philadelphia following consolidation that the avenue appeared (see Consolidation).

In the late nineteenth century, several deeds transferred land for use as an avenue—mainly from 31st to 34th Street and from 41st Street to Belmont Avenue. It was not until 1890 that the section from 33rd to 35th Street was confirmed. Two years later, the road was extended to complete the gap, and the avenue achieved its full distance from 31st Street to Belmont.

Maple Street

See Trenton Avenue.

Maplewood Street

See Armat Street.

Margaretta Street

See Sartain Street.

Mark's Lane

See Quarry Street.

Market Street

William Penn decreed that the main east-west street of his new city would be called High. Regardless of the founder's wish, the thoroughfare was, by 1750, commonly called Market. (When another marketing street grew up in Philadelphia, it was, naturally, named New Market Street, to distinguish it from the "old" one.)

"The market house," an 1824 traveler wrote, "which is nothing more than a roof supported by pillars and quite open to each side, begins on the bank of the Delaware, and runs one mile, that is, eight squares in length!" In the middle of the market place, to the east of 2nd Street, stood the prison, pillory, and stocks. The prison was removed in 1722. Shortly thereafter, at the instigation of Benjamin Franklin, the first permanent street paving in the city was installed along the route.

The earliest survey of lots on Market or High Street was recorded in 1682 on the west side of the street between 3rd and 4th Streets. The name of the street was officially changed to Market in 1853 to conform to common use. By 1883, Market Street was declared a public street from the Delaware to the Schuylkill.

West of the Schuylkill, the street was confirmed as far as Cobbs Creek in 1788. The section which ran through Hamilton Village in West Philadelphia was known as Washington Street prior to 1858. The entire

MARKET STREET. Originally named High, this street's name was changed to reflect its popular use: as a marketplace. Shown is an 1838 view of "The Jersey Market," Market Street at Front.

PENN CENTER. Market Street at 15th, as it looked in 1941. The structure that appears in the background was Philadelphia's famous "Chinese Wall." The elevated wall carried Pennsylvania Railroad trains. After the wall was razed, the railroad used underground tracks to get trains from 30th Street Station to downtown (Suburban Station). (Philadelphia City Planning Commission)

stretch of street was widened to its present width between 1870 and 1931.

Markoe Street

Markoe Street is named for Abraham Markoe (1727–1806), founder of the first volunteer military association in what is now the United States: the Philadelphia Light Horse, now known as the First City Troop. (See Fitzwater Street.)

Marvine Street

See Sartain Street.

Mary Street

See League Street.

Mascher Street

See House Numbering, Letitia Street.

McAllister Street

See Quarry Street.

McClellan Street

George Brinton McClellan (1826–85) was a Philadelphian who entered West Point at 15, and graduated second in his 1846 class.

"Little Mac" showed great promise in the early hours of the Civil War. In the Department of the Ohio, he secured what is now West Virginia by July 1861. He then took command of the Army of the Potomac. There, McClellan was overly cautious, and reluctant to do battle with the Confederates. In November 1862, he was relieved of command. Two years later, the Democratic Party nominated him for president on a peace platform. The Democrats were soundly defeated, and McClellan resigned his commission and sailed to Europe for a three-year visit.

McCrea Street

See Fulton Street.

McIllery Street

See Sartain Street.

McKean Parkway

See McKean Street.

McKean Street

Though some authorities claim McKean Street was named for the Pennsylvania county, it probably, because of its location, was named for Thomas McKean, an early Pennsylvania governor (see Governors). The choice is less than crucial, however, since the county was also named for McKean (1734–1817), who was the only signer of the Declaration of Independence to sit in the Continental Congress from the beginning to the end of the Revolution.

Originally called the McKean Parkway, the roadway was first confirmed, from Moyamensing to Broad, in 1873. Before the turn of the century, McKean had extended from 26th Street to Delaware Avenue. The remaining several blocks were added in the 1950s.

McKEAN STREET. Thomas McKean, a former governor of Philadelphia, was the only signer of the Declaration of Independence to sit in the Continental Congress from the start to the finish of the Revolution. Along with John Dickinson, he was the only other congressman to serve in the Continental Army. (Philadelphia Municipal Archives)

McKinley Street

McKinley Street was named to honor the memory of William McKinley (1843–1901), twenty-fifth president of the United States.

McKinley served in an Ohio regiment during the Civil War, as aide to Colonel Rutherford B. Hayes. At the end of the war, he had been promoted to major. Returning to Ohio, he studied law, but soon became involved in politics. Except for one term, McKinley served in the U. S. House of Representatives from 1876 to 1891. He was elected governor of Ohio in 1891 and 1893. He was elected president in 1897, only to face continuing problems in Cuba, which erupted into the Spanish-American War with the sinking of the battleship *Maine* in Havana harbor.

Re-elected in 1900, McKinley was shot by Leon Czolgosz, an anarchist, while visiting the Pan-American Exposition in Buffalo, New York. He was succeeded by his vice-president, Theodore Roosevelt.

McMahon Avenue

Named for the man who developed virtually all of East

McMahon Avenue

Germantown, McMahon Avenue came into being in 1907, when the city legislated to change the name of Underhill Street, from Mechanic Street to Locust Street, to McMahon. Underhill had existed, by deed, from 1888.

David McMahon (1831–1912) was a Philadelphia contractor who built many homes in the area and deeded the roadbeds to the city. He was also a charitable man who helped the poor and needy in Germantown.

The last addition to McMahon Avenue, from East Walnut Lane to Tulpehocken Street, was made in 1922.

McMichael Street

McMICHAEL STREET. Morton McMichael, for whom this street is named, was publisher of the *North American* newspaper, first president of the Fairmount Park commissioners in 1867, and mayor of Philadelphia. (Scharf and Westcott's *History of Philadelphia, 1609–1884*)

Morton McMichael (1807–79), for whom this street was named, was publisher of the *North American* newspaper, mayor of Philadelphia, and the first president of the Fairmount Park commissioners in 1867. A bronze statue of McMichael stands in the park, inscribed "An honored and beloved citizen of Philadelphia."

The section of McMichael Street from Midvale to Fairview was dedicated in 1889. The street, which once included a section from Hunting Park Avenue to Roberts (vacated in 1953), extends from Roberts to Abbottsford and from Queen Lane to Coulter.

McNulty Road

Black Lake Road, from a dead end west of the Southampton-Byberry Road to Meeting House Road, became McNulty Road in 1984. The road was named in honor of Thomas J. "Reds" McNulty, business manager of Local 690 of the Plumbers Union from 1972–1982. McNulty, who died in 1983, stepped down as union boss to become business manager emeritus, and was eulogized as "a plumber first, and a man who believed in people and believed in the labor movement. . . ."

Mead Street

See Fitzwater Street.

Meade Street

General George Gordon Meade (1815–72), commander-in-chief of the Army of the Potomac during the Civil War, was honored by the city of his childhood in many ways. "After the Civil War," a newspaper reported, "Philadelphians bought him a house . . . and following his death, collected $100,000 for his heirs." It was in keeping that such a grateful city should name a street in his honor.

On the other hand, his grandfather, George (1741–1808), was a native Philadelphian and one of the incorporators, along with Mathew Carey (see Carey Street), of the Hibernian Society in 1792.

In any case, Meade Street, from Navajo to Shawnee, was open as early as 1905. The section from Ardleigh to Anderson was opened by deed in 1906.

MEADE STREET. General George Gordon Meade, Civil War hero, donated his name to the Philadelphia street, and also to Fort George Gordon Meade, Maryland. (Scharf and Wescott's *History of Philadelphia, 1609–1884*)

Mechanic Lane

See Roxborough Avenue.

Mechanic Street

See League Street.

Mechanicsville Road

Leading to an early village by that name, Mechanicsville Road was confirmed by Quarter Sessions Court in 1832. Running from Nanton Drive to Knights Road, it was widened between 1954 and 1962.

Medary Avenue

William F. Medary owned land in the vicinity of Church Lane and Old York Road during the late nineteenth century. It was in recognition of him and his family that this avenue was named.

Medary Avenue

The Medary family in Philadelphia dates back to 1739, when Jacob Madery arrived here. His son Sebastian altered the spelling of the family's name.

Medary Avenue was deeded to the city, from Broad to 12th, part of William's land, in 1897. By 1936, the avenue extended to its present length—from Wister to Stenton; Old York Road to 15th; Broad to 10th; and 8th to Lawrence.

Meeting House Lane

See Haines Street.

Mercer Street

See Armstrong Street, Counties, Rubicam Street.

Mermaid Lane

Mermaid Lane, first confirmed in 1804 from Stenton Avenue to Germantown, commemorates the Mermaid Hotel. Dating back to 1795, the hotel was a favorite spot for eighteenth-century turkey shoots. The remainder of this roadway was deeded to the city between 1887 and 1920. Prior to 1900, Mermaid was known as an avenue from Stenton Avenue to McCallum Street.

Mershon Street

Dedicated in the city in 1926 from Devereaux Avenue to Levick Street, Mershon Street presumably honors a member of the Mershon family. The most famous Mershons were Cornelius and George, who operated the Mershon Patent Shaking Grate Works in 1915. At the same time, however, the Mershon brothers (Charles O. and Abner H.) were active real estate operators. It is possible that this street commemorates the realtors.

Methodist Lane

See Haines Street.

Midvale Street

See Edgley Street, Ruffner Street.

Mifflin Street

Named for General—later Governor—Thomas Mifflin, this street was first recorded in public use from Chadwick to Bancroft Street in 1855. Less than twenty years later, the jury confirmed the distance from 6th to 9th. Three years later, in 1874, another piece was deeded to the city—from 16th to Broad.

Other sections of Mifflin Street were in use during that time—principally Passyunk to Juniper and Vandalia to 12th—but not officially. It was not until the end of the nineteenth century that those portions of Mifflin were officially opened.

From Delaware Avenue to Point Breeze, the street was complete and in public use by 1899. West of 22nd Street it was completed to 35th Street in 1939.

Thomas Mifflin, first governor of the Commonwealth under the 1790 Constitution, was quartermaster-general under Washington during the Revolution. The street to carry his name is the first in the line of streets bearing the names of Pennsylvania governors (see Governors). Mifflin also lent his name to Fort Mifflin, "the Alamo of the Revolution," in southwest Philadelphia. The approach to the fort is called Fort Mifflin Road.

MIFFLIN STREET. As a general in Washington's army and later as governor of Pennsylvania, Thomas Mifflin contributed greatly to Philadelphia's progress. (Engraving "from the original painting in the possession of Alex. J. Dallas Dixon, Esq., Phila.")

Milbourne Street

See Bancroft Street.

Milden Hall Street

See Bouvier Street.

Miller Street

See Gowen Avenue, Willow Street.

Minerva Street

See Brandywine Street.

Mint Court

See Quarry Street.

Mission Street

See Bonsall Street.

Mohican Street

See Indian Tribes.

Monastery Avenue

The leader of the Seventh-Day Baptists, Joseph Gorgas (see Gorgas Lane) bought ground on the Wissahickon in 1752 to construct a three-story stone house, still standing and known as the Monastery.

Tradition has it the house was used for seclusion and religious meditation. Gorgas's monks, it is said, wore long white robes with cowling, much like that of the Capuchins. Gorgas sold the house and the ground

Montgomery Avenue

in 1761 to Edward Milner. It hasn't been used for monkish purposes since.

However, the house and the name persist. The original road which led to the Monastery existed much earlier than 1865—even if not as a legally open street. The name was given to the avenue in 1906, when Levering Avenue became Monastery Avenue. The roadway which led to Gorgas's monastery has since been stricken from the city plan, and the avenue goes only from Henry Avenue to Boone Street.

MONASTERY AVENUE. Joseph Gorgas, leader of the Seventh-Day Baptists, used this monastery for seclusion and meditation. The Monastery, still standing along the Wissahickon near Kitchens Lane, was built in 1752.

Monroe Street

See Counties.

Montgomery Avenue

See Counties.

Moore Street

See Fletcher Street.

Moravian Street

The first Moravian Church in Philadelphia was built in 1742 on Franklin Street (now Girard Avenue) at the corner of Wood Street. The alley on which the church stood was named Moravian Alley. Though the alley's name was later changed to Bread Street, the contribution of the Moravians is remembered by this street, which runs—with frequent interruptions—from Jessup to 21st.

The Moravians didn't grow as rapidly as did other denominations which came to Philadelphia. Although they were the oldest Protestant Church, founded in 1457, and had sent missionary excursions to the colonies as early as 1735, their main thrust was only felt in those communities which they either founded or settled, like Bethlehem and Nazareth, Pennsylvania.

Bethlehem was named by Count Nicholas Zinzendorf, who expected the town to become the center of Moravian conversion of the American Indians. Instead, it soon became the site for an asylum, school and academy. Nazareth, on the other hand, was purchased

MORAVIAN STREET. Members of Philadelphia's early 18th century Lutheran Church were wooed from their faith by Count Nicholas Louis Zinzindorf, founder of the Moravian Church.

by the Moravians in 1743. With it, they obtained the rights of a court baron . . . the only manor which the Penns transferred with such privileges. The annual payment for this feudal right was the presentation of a single red rose each June.

According to official city records, the oldest section of this street to bear the name Moravian is from Broad Street to 15th, confirmed in 1837. In 1895, through the name changes of Gold (Dock Street to 2nd, vacated in 1958), Lawson (Jessup to 12th), Lewis (16th to 18th), Porcelain (19th to 21st), and Lewis (36th to McAlpin, vacated in 1968), Moravian Street achieved its present configuration.

Morris Street

See Spring Garden Street.

Mount Airy Avenue

Taking its name from the village of Mount Airy, formerly called Cresheim (see Cresheim Road), Mount Airy Avenue was the main road north of Germantown.

The village of Mount Airy, considerably higher in elevation and separated from Germantown, presumably takes its name from the pre-Revolutionary War country-seat of Chief Justice William Allen. Another source implies that the village obtained its name from "the airy position of the district." It is more likely that Judge Allen used such a criterion in naming his seat and the village later assumed the same name.

Part of the original Willow Grove and Germantown Plank Road, Mount Airy Avenue was begun in 1853 from Germantown Avenue to Willow Grove. It was freed from toll in 1890 (see Toll Roads). Prior to 1941, the avenue was known as a street. Including a portion of Gorgas Mill Road (confirmed in 1764), Mount Airy Avenue extends from the line of Fairmount Park by Wissahickon Avenue to the county line.

Moyamensing Avenue

Modern-day Moyamensing Avenue extends from Penrose Ferry Road to Christian Street. Its name is taken from an early village located to the south of the city. Olaf Stille, one of the original Swedish settlers, acquired the land, mostly swamp, as early as 1678. Maps of the city drawn in that year show the line of this road.

The Moyamensing Road was put into official use in 1790 from the line of the District of Southwark to the Penrose ferry. In 1858, Moyamensing was extended by the incorporation of old Jefferson Avenue from Mifflin to Christian. It was probably at that time that the road became an avenue. The last segment to be added—from Snyder Avenue to Mifflin Street—was confirmed in 1887.

The interpretation of the word *moyamensing* is disputed. It has been translated to mean "a place of meeting," "the place for maize," "an unclean place," and "pigeon roost." In the Lenni-Lenape language, the name should be translated as "pigeon excrement."

MOYAMENSING AVENUE. The avenue received its name from the township in which it was a major route. A major sight for many years on Moyamensing Avenue was Moyamensing Prison. A military execution at Fort Mifflin during the Civil War even used a gallows borrowed from this prison. According to the Lenni-Lenape language, *moyamensing* means "pigeon excrement." Does the definition have anything to do with "stool pigeons"? (Scharf and Westcott's *History of Philadelphia, 1609–1884*)

Mud Lane

See Paschall Avenue.

Mulberry Street

See Arch Street, Camac Street.

Mulvaney Street

See Camac Street.

Murray Street

See Rittenhouse Square.

N

Narragansett Street

See Indian Tribes.

Navajo (Navahoe) Street

See Indian Tribes.

Naval Asylum Place

See Fulton Street.

Neff Street

See Trenton Avenue.

New Hickory Lane

See Fairmount Avenue.

New Market Street

See Market Street.

New Street

See Callowhill Street.

Nice Street

See Johnson Street.

Nicetown Lane

Nicetown Lane, named for an early Philadelphia village, was first confirmed by the road jury in 1789, from Frankford to Germantown. Improved between 1923 and 1926, the lane now officially extends from Butler Street to Front. Additional sections, from Front to Luzerne and from G Street to Palmetto, were vacated between 1930 and 1944.

The village of Nicetown was named for the Nice or Neus family, owners of Cedar Grove, a country estate in Nicetown (see also Hunting Park Avenue).

Nippon Street

See Allens Lane.

Nixon Street

See Flat Rock Road.

Noble Street

There is a popular legend that Noble Street was originally called Bloody Alley because a murder had been committed there. In an effort to remove the stigma, the tale goes, the name was changed to Noble.

Unfortunately, there is little information to prove or disprove this contention. The most useful fact that can be found is that there was a Noble family who came to Philadelphia from Bristol, England, in 1684. The fact that they arrived with the founding fathers might have been sufficient to have a street named in their honor. But, as with the folk tale, there is little or nothing to prove it.

Noble Street first appeared on record when an act was passed in 1768 ordering that £2,000 be raised to purchase land between the Frankford Road and the Delaware River. The money was raised, and later that same year Philotesia Strettell conveyed the land to the

NIXON STREET. John Nixon, Philadelphia's High Sheriff, was the person selected to make the first public reading of the Declaration of Independence. He read it from the steps of the State House . . . after members of the Continental Congress were safely out of town. (Scharf and Westcott's *History of Philadelphia, 1609–1884*)

county commissioners of Northern Liberties. A road from Front Street to Beach was confirmed in 1808.

In 1819, another section appeared—from 6th to 9th. This was vacated in 1967. By 1883, Noble included the blocks from 9th to New Market. Hoffman (62nd to 61st), Tatlow (19th to 18th), and James (Broad to 9th) were added by a change of name in 1895. After having several sections vacated, Noble presently extends from 62nd to 61st, from Broad to Lawrence, from 3rd to New Market, and from Front to Delaware Avenue.

Norris Street

Norris Street was named for Isaac Norris, a wealthy merchant and diplomat who left London for Philadelphia in 1693. He served an active civic life as a member of the Provincial Council, speaker of the Assembly, and mayor of Philadelphia.

From the Delaware River to 2nd Street, Norris Street was first confirmed in 1840, through land once owned by the Norris family. With the introduction of Lancaster Street—the name was changed to Norris in 1858—the street gained length to the west, from Stillman to Broad. By 1870, Norris had been extended on the east as far west as Germantown Avenue. A few years later, the east and west sections met. Norris Street from 33rd to 25th Street was acquired by deed before the end of the nineteenth century.

Norton Street

See Cliveden Street.

Numbered Streets

Since the time of William Penn, Philadelphia has had a system of numbered streets (see Original Streets). At present, they run north and south from 2nd Street to 88th. The northern reaches of the city also contain avenues numbered 64th to 80th.

Oak Street

See Beach Street.

Ogontz Avenue

An Indian chief who entertained a young boy—with a long memory—lent his name to this major Philadelphia thoroughfare. Jay Cooke, a financier of the Civil War, named his Oak Lane mansion in 1865 for Chief Ogontz, who had entertained him and his friends at the Cooke family's home in Sandusky, Ohio. Cooke lost the mansion and the estate, Chelton Hill, in the "panic of 1873," but regained his property in 1881.

The earliest extent of Ogontz Avenue was from Ruscomb Street to Lindley Avenue, deeded to the city in 1885. The remainder, from Somerville to Cheltenham Avenue, was deeded and put in use before 1930.

Old Meeting House Road

See Verree Road.

Old York Road

See Toll Roads.

Original Streets

After his first visit to "the virgin settlement of [his] Province," William Penn wrote: "The city consists of a large Front Street to each river, and a High Street (near the middle) from Front (or River) to Front, of 100 foot broad, and a Broad Street in the middle of the city, from side to side, of the like breadth . . . and eight streets (besides the High Street) that runs from Front to Front, and twenty streets (besides the Broad Street) that run

Original Streets

across the city, from side to side; all these streets are fifty feet breadth."

As late as 1720, however, there were but four streets running parallel with the Delaware River. During the Revolution, "the town extended only from Christian to Callowhill Streets, north and south, and houses built as far west as Tenth Street might fairly be classed as country seats."

Insofar as they exist today, Penn's *designated* north-south streets—those he intended to become streets—are Vine Street (formerly Valley), Race Street (formerly Sassafras and Songhurst), Arch Street (formerly Mulberry and Holme), Market Street (formerly High), Chestnut Street (formerly Wynne), Walnut Street (formerly Pool), Spruce Street, Pine Street, and South Street (formerly Cedar). From east to west, numbered streets began at the Delaware River and were numbered as Delaware Front, Delaware 2nd, and so on to Broad Street. To the west, the streets began at the Schuylkill River and were listed as Schuylkill Front, Schuylkill 2nd, and so forth back to Broad. The streets are now listed consecutively beginning at the Delaware.

The growth and development of Penn's original streets—those which were in use—can be deduced from the following listing of lot surveys located in Nevell's "Extracts."

Extent	Survey	
	Earliest	*Latest*
Front Street—*East Side*		
Pine to Spruce	October 2, 1685	
Spruce to Walnut	August 6, 1684	July 12, 1690
Walnut to Chestnut	April 14, 1684	June 11, 1689
Chestnut to Market	January 26, 1689/90	March 1, 1701
Market to Arch	April 3, 1689	January 30, 1689/90
Arch to Race	March 15, 1689	January 23, 1693/94
Race to Vine	March 30, 1689	March 2, 1740
Vine and above	March 24, 1689	
Front Street—*West Side*		
Cedar to Pine	August 1, 1684	
Pine to Spruce	July 25, 1684	October 2, 1685
Spruce to Dock—"granted by governor Penn to the city and (Richard) Marsh went higher up for his lot"		
Dock to Walnut	January 30, 1683/84	June 7, 1688
Walnut to Chestnut	April 3, 1682	February 17, 1684
Chestnut to Market	April 4, 1683	September 6, 1692

Original Streets

Extent	Survey	
	Earliest	*Latest*
Market to Arch	April 7, 1683	July 5, 1717
Arch to Race	June 21, 1683	August 24, 1703
Race to Vine	August 10, 1682	December 20, 1704
Second Street—*West Side*		
Cedar to Spruce	August 1, 1684	October 2, 1695
Spruce to Walnut	August 1, 1642*	October 24, 1701
Walnut to Chestnut	February 28, 1682/83	December 17, 1683
Chestnut to Market	January 12, 1683/84	October 17, 1689
Market to Arch	September 15, 1683	October 12, 1702
Arch to Race	June 26, 1683	May 14, 1702
Race to Vine	June 12, 1683	January 13, 1708/09
Third Street—*East Side*		
to Arch	August 1, 1684	November 16, 1717
Third Street—*West Side*		
to Race	June 12, 1684	March 29, 1706
Race to Vine	February 5, 1705/6	December 28, 1730
Strawberry Alley—*West Side*		
	June 10, 1690	November 24, 1692
Walnut Street—*North Side*		
Dock to 3rd	October 11, 1684	February 29, 1704/5
3rd to 4th	April 27, 1683	October 31, 1691
4th to 5th	March 27, 1683	October 31, 1691
5th to 6th	November 22, 1683	October 26, 1715
6th to 7th	April 3, 1716	
Walnut Street—*South Side*		
Dock to 3rd	October 13, 1682	June 25, 1688
3rd to 4th	February 29, 1674/75	June 20, 1688
4th to 5th	August 3, 1684	December 26, 1691
5th to 6th	August 15, 1634*	June 6, 1684
Market Street—*North Side*		
2nd to 3rd	August 2, 1684	October 25, 1704
3rd to 4th	January 29, 1682/83	January 12, 1702/03
4th to 5th	June 13, 1684	September 4, 1707
5th to 6th	June 26, 1683	June 10, 1718
6th to 7th	October 8, 1684	August 22, 1706

---

*This apparently is an error in transcribing from the original surveys.

Original Streets

Extent	Survey	
	Earliest	*Latest*
Market Street—*South Side*		
2nd to 3rd	November 15, 1683	June 14, 1690
3rd to 4th	September 24, 1684	July 16, 1705
4th to 5th	December 21, 1683	November 9, 1722
5th to 6th	August 19, 1683	October 2, 1704
6th to 7th	June 19, 1683	September 18, 1701
7th to 8th	August 7, 1685	April 26, 1687
Race Street—*South Side*		
to 4th	March 27, 1704	December 2, 1707
4th to 5th	October 5, 1686	July 20, 1717
5th to 6th	May 6, 1714	
Arch Street—*North Side*		
3rd to 4th	May 20, 1683	December 26, 1684
4th to 5th	March 26, 1701	October 10, 1717
5th to 6th	August 27, 1684	August 1, 1707
6th to 7th	July 12, 1718	April 10, 1738
7th to 8th	June 25, 1690	March 3, 1740
Arch Street—*South Side*		
3rd to 4th	December 6, 1683	February 28, 1700/1701
4th to 6th	no record	
6th to 7th	August 17, 1699	
7th to 8th	December 6, 1690	

These surveys give a good indication of the dates of land appropriation. Unfortunately, they do not cover all the streets which were supposedly in use at the time. Perhaps, on the other hand, these were the only sections of the streets which were being used up until 1740. No one knows for sure.

Orion Road

Orion Road, running from Vinton to Medford Road, recalls an old property, the Orion Tract. Originally purchased in 1671, the tract was owned by William Orion, a blacksmith "and a man of authority in Upland," England. Orion's estate was called Edgley (see Edgley Avenue). Present-day Orion was deeded to the city for public use in 1959.

Orthodox Street

In 1834, a road was confirmed from Frankford to Leiper. Its route passed the meeting house of the Orthodox Friends at the corner of Penn Street. It is from this old house of meeting that the street obtained its name.

Orthodox Street, by 1856, extended from Leiper to the Frankford Creek. Less than twenty years later, the street was opened—to the west, to Adams Avenue; to the east, to Richmond Street. Not until the dawn of the twentieth century did Orthodox achieve its full distance from Castor to Carbon Street.

The Orthodox Friends were the group that adhered to the original, evangelical doctrines of the Society of Friends when the more liberal group left after the Hicksite split.

Osage Avenue

See Indian Tribes.

Otsego Street

See Water Street.

Overbrook Avenue

Overbrook Avenue takes its name from the section of Philadelphia by the same name. The community, in turn, obtained its name from the fact that railroad tracks at the point of junction between the community and Montgomery County crossed "over a brook." The tract of land, from 58th to 66th, and from City Line to Woodbine, was purchased by Drexel and Company after 1887 and developed as Overbrook Farms.

The earliest section of the avenue was deeded to the city in 1893—from 54th Street to Bryn Mawr Avenue. In 1899, the street appeared from 64th Street to Cardinal Avenue. Completing Overbrook to its present length—

Overbrook Avenue

Ashurst to Haverford (in a new location), 66th to 50th, and 46th to Belmont Avenue—the last deed was presented in 1925.

Owen Street

See Bucknell Street.

Oxford Avenue

See Toll Roads.

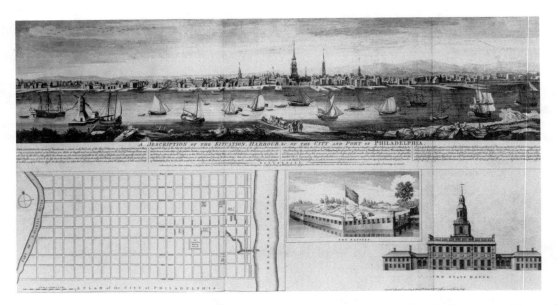

A DESCRIPTION OF THE SITUATION, HARBOUR &C. OF THE CITY AND PORT OF PHILADELPHIA. Undated English sketch of Philadelphia and its environs. (U.S. Office of War Information, National Archives)

Oxford Avenue

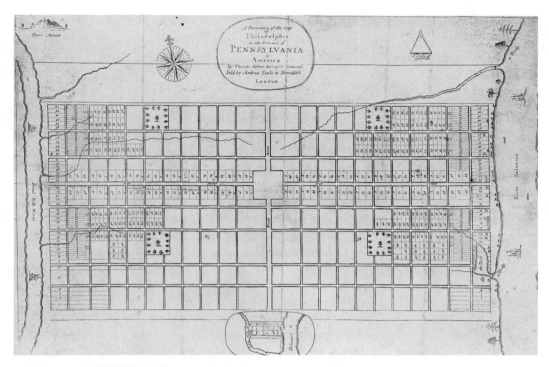

PENN'S PLAN FOR PHILADELPHIA. Thomas Holme, Penn's surveyer-general, drew up this 17th century "Portraiture of the City" to help promote land investment. The city did not look at all like Holme's plan until much, much later. (Philadelphia City Planning Commission)

Packer Avenue

Built as an approach to the "exposition ground of the Sesqui-Centennial" of 1926, Packer Avenue commemorates a small-town newspaper publisher who became Pennsylvania's chief officer.

William Fisher Packer (1807–70) learned the newspaper business as a journeyman printer on the Sunbury (Pa.) *Public Inquirer,* the *Bellefonte Patriot,* and the *Pennsylvania Intelligencer,* owned at the time by Simon Cameron (see Cameron Street), the "Czar of Pennsylvania Politics." In 1827, he became connected with the *Lycoming Gazette.* Two years later, at age twenty-two, he was the sole owner of the paper. Using his newspaper as a sounding board, Packer became a leader in Democratic circles.

Packer served two terms as a state representative and one as a senator. In 1857, he was elected governor (see Governors). During his tenure, he pushed for state aid for the construction of the Sunbury & Erie Railroad, a measure which passed after he left office.

Originally built from Moyamensing Avenue to Broad Street, Packer Avenue now goes from 24th to Swanson. The avenue has been widened, mainly in 1957, and a portion, from Lawrence to 10th, exists in the bed of former Curtin Street (see Curtin Street).

Pallas Street

See Camac Street.

Palmer Street

See Kensington Avenue.

Palmer's Lane

See Queen Lane.

Palumbo Plaza

In 1975, Philadelphia's City Council decided to honor Frank Palumbo, owner of South Philadelphia's landmark restaurant and nightclub, by naming Darien Street, the strip of roadway that separated the night club from its parking lot, Palumbo Plaza.

Palumbo's had been the entertainment center of South Philadelphia since 1884, when Frank Palumbo's father, Antonio, opened it as a boarding house for Italian immigrants. When the Italians landed at Philadelphia's Washington Avenue wharf, they wore signs that read "Palumbo's," and that's where they would go . . . to the rooming house on Fallon Street, between Catharine and Christian. Once at Palumbo's, they were greeted by strolling musicians and an Italian meal. Relatives would come to Palumbo's to pick up their kin. Fallon Street was later renamed Darien.

Frank Palumbo, known for his generosity to many charities, including the Philadelphia Zoo, brought many of the entertainment industry's biggest names to his establishment, including Frank Sinatra, Jimmy Durante, Rosemary Clooney, Frankie Laine, Louis Prima, and others. Because of the entertainment, or despite it, Palumbo's was a gathering spot for Philadelphia's politicos. During any election campaign, the restaurant would be filled almost nightly with campaigners, groupies, and hangers-on. Palumbo died in 1983 at the age of 72. Six years later, Palumbo Plaza was no longer. Its name returned to Darien Street. Rumor had it that some people felt Darien Street was too insignificant, and unworthy of the Palumbo name.

In 1987, members of Philadelphia's City Council decided to approve a bill that would rename West River Drive in Palumbo's honor. The council's interest in the Palumbo name change was sparked by Philadelphia radio personality Irv Homer. Homer was able to marshal a force of 30,000 citizens, who sent letters and postcards. Philadelphia's lawmakers, like those elsewhere, do bend a bit to public pressure, even if it's not their area of responsibility.

Perhaps because it wasn't their area that the council made its vain gesture. Philadelphia's legal counsel opined that the authority of the naming of streets in Fairmount Park rested with the Fairmount Park Commission—not City Council. Citing an 1867 law, First Deputy City Soliciter Brian Appel wrote, "maintenance of the character of the park" is delegated to the commission. "Inherent in this power to determine" led the commission to rename East River Drive as Kelly Drive shortly after the untimely death of John B. Kelly, Jr., in 1985 (See Kelly Drive).

At the Fairmount Park Commission, where the buck stopped, commissioners voted 11–2 not to rename the drive, after a few words of praise were said in Palumbo's honor.

"Maybe in retrospect," Commission President F. Eugene Dixon said, "we acted too hastily in renaming Kelly Drive" within weeks after Kelly's death. The commission, Dixon added, "has no desire to rename anything." During the summer of 1986, the commission adopted a policy that it would not rename anything else in the park for anyone dead fewer than ten years.

Kippee Palumbo, the restauranteur's widow, attended the meeting and was disappointed in the decision. "At the same time," she said, "I feel a lot of love for the 30,000 people who wrote cards and letters and for the many important people who went on the air to tell of Frank Palumbo's good works." As of this writing, the matter is still not settled.

Park Street

See Wishart Street.

Parrish Street

A staunch abolitionist doctor, who witnessed the 1833 deathbed will of John Randolph which freed the Virginian's slaves, was honored by his native city in the naming of Parrish Street.

Joseph Parrish (1779–1840), a descendent of Lord Baltimore's surveyor-general, studied medicine under

Caspar Wistar (see Wistaria Street). He received his medical degree from the University of Pennsylvania in 1805. During his career, Dr. Parrish was a physician to the Philadelphia Almshouse (1807–11) and later surgeon there (1811–21), a member of the staff of the Pennsylvania Hospital (1816–29), president of the board of managers of the Wills Eye Hospital (1833–40), and vice-president of the College of Physicians of Philadelphia and the Philadelphia Medical Society. In addition to hating slavery, Parrish also advocated the abolition of the death penalty.

Parrish Street appears on plans of the District of Spring Garden as early as 1834. These show a tract of land along present-day Parrish east and west of Marshall Street as being owned by Joseph Parrish. The earliest record of such a street is in a jury confirmation of 1833, from 13th Street to Broad. After the doctor's death in 1840, the street grew to its present length and incorporated School (Leithgow to Lawrence), Valeria (16th to Francis), and Seneca (Lancaster Avenue to 50th Street) in 1895. The last portion of Parrish was incorporated in 1904, from 50th Street to 51st.

Paschall Avenue

Paschallville was settled in 1682 by Thomas Paschall on a five-hundred-acre grant of land east of Cobbs Creek. The road which runs through the old village was named Pascall.

The earliest recording of Paschall Avenue, dated 1848, is from Island Avenue to Grovers Avenue (formerly Mud Lane). Completed by 1927, Paschall Avenue extends from 46th Street to Island.

Passyunk Avenue

The land between the Delaware and Schuylkill rivers, near the mouth of the Schuylkill, was called by the Indians *pachsegink* or *pachsegonk*, meaning a level place, a valley, or a place between the hills. The name was

Passyunk Avenue

maintained by the Swedes who settled there in 1648 and later by the English. Passyunk grew from a small village to a township, which was consolidated (see Consolidation) with the city of Philadelphia in 1854.

The road through that section, confirmed in 1750 from South Street to the Schuylkill River, was called the Passyunk Road. It followed an earlier Indian footpath. In 1870, the road became an avenue. The last recorded deed for the expansion of Passyunk Avenue was in 1943. It now extends from Elmwood and Island Avenues to South Street.

On early city records, Passyunk can be found spelled Passajon, Passajungh, Paisajung, Perslajungh, Passuming, Passyonck, and Passayunk. Regardless of varied spellings, South Philadelphians still pronounce it: Pash-unk.

Pastorius Street

See Germantown Avenue.

PASTORIUS STREET. Germantown's founding father, Francis Daniel Pastorius, once lived in this house. The leader of the Crefelders, early immigrants from Crefeld, Germany, Pastorius had houses built on Main Street—now Germantown Avenue—facing each other: for mutual protection. (Scharf and Westcott's *History of Philadelphia, 1609–1884*)

Pattison Avenue

See Governors.

Paul Street

See League Street.

Peale Street

See Belfield Avenue.

Pear Street

See Warren Street, "Things that Grow . . ."

Pegg Street

See Willow Street.

Pelle Circle

See Pelle Road.

Pelle Road

Pelle Lindbergh was a goalie for the Philadelphia Flyers who died in 1985 in a car crash. Pelle Circle and Road in Philadelphia's Northeast were dedicated in his honor in 1986.

A player noted for his saves, Pelle "died making another save," as sportswriter Mike Keenan said in his obituary. Pelle Lindbergh's organs were donated for transplant use. "Pelle Lindbergh," Keenan wrote, "was quick of glove and legs, and lighthearted in spirit. He played goal by the book and speed-read through life."

Lindbergh, it must be noted, had blood-alcohol levels of 0.17 and 0.24 percent, well above the legal limit of 0.10. Police said that Lindbergh had been speeding along in his high-powered Porsche when he missed a turn and crashed into a concrete wall beside an elementary school in Somerdale Borough, New Jersey.

As a result of the attention drawn to this streetnaming, Philadelphia developer Fredavid Greenberg decided to name streets in his Northeast developments after Philadelphia police officers and firefighters who died in the line of duty. The first such street, Joseph Kelly Terrace, was dedicated November 15, 1988, in honor of Joseph Kelly, a 16-year veteran who was shot to death in 1971 while apprehending a suspected car thief. Greenburg promised to continue the practice in other Fredavid developments, including the Mansion of Timberwalk, Bentley, Liberty Square, and Liberty Circle.

This roadway would have been named "Lindbergh," but the City of Philadelphia had another by that name (see Lindbergh Boulevard).

Penn Square

See Callowhill Street, Squares.

Penn Street

See Callowhill Street, Water Street.

Pennsgrove Street

See Callowhill Street.

Pennsylvania Avenue

See Callowhill Street, Kennedy Boulevard.

PENN SQUARE, PENN STREET, ETC. William Penn's head usually sits atop Philadelphia's City Hall. This photograph, taken during the construction of the tower, provides a view of the city's founder that can't be duplicated today. (Philadelphia City Planning Commission)

Phil-Ellena Street

The name of this Germantown street is truly remarkable. It stands as a symbol of the love and affection George W. Carpenter held for his wife, Ellen.

Between 1841 and 1844, Carpenter built a mansion on his five-hundred-acre estate in Germantown. The house, torn down in the 1890s, stood at what is now 5510 Germantown Avenue. To apply a fitting name to the house, Carpenter adapted the Greek word for love combined with his wife's first name. The result was Phil-Ellena.

Opened by affidavit, between 1884 and 1910, from Germantown Avenue to Chew, Phil-Ellena was first known as Church Street until 1895, as far as Cheltenham Avenue. The remaining sections of this street were deeded to the city between 1899 and 1929.

Philip Street

See Strawberry Street.

Pickius Lane

See Haines Street, Manheim Street.

Pike Street

See Counties, Iseminger Street.

Pine Street

See Original Streets.

Plumstead Lane

See Fairmount Avenue.

Plymouth Street

See Rittenhouse Street.

Point Breeze Road

See Schuylkill Avenue.

Pollock Street

Once stretching from Front Street to Moyamensing Avenue, Pollock Street has shrunk to its present length as a result of legislation in 1947, 1956, 1960, and 1964. The section of Pollock which once ran from 7th to Darien Street is now Stella Maris. There is also, to add confusion, a Pollock Terrace.

James Pollock (1810–90), for whom this street was named, was a Pennsylvania governor (see Governors) and a descendant of an early Pennsylvania settler. Pollock served in the U.S. House of Representatives from 1844 to 1849. He was one of the few men who encouraged telegrapher Samuel F. B. Morse when the inventor traveled to Washington to secure governmental assistance for his project.

Nominated by both the Whig and the Native American ("Know-Nothing") parties, Pollock defeated William Bigler (see Bigler Street) for the governorship, and served from 1855 to 1858. The chief accomplishment of his administration was the sale of the state-owned canals to several railroads, principally the Pennsylvania Railroad. This move reduced the state debt and ultimately lowered taxes.

Pollock refused renomination and returned to his law practice. Three years later, he was appointed director of the U.S. Mint in Philadelphia. During his appointment (1861–66), he suggested that "In God We Trust" be placed on all United States currency large enough to contain it. The practice is continued to this day.

Pollock Terrace

See Pollock Street.

Pool Street

See Walnut Street.

Poor House Lane

See Rittenhouse Street.

Poquessing Street

See Torresdale Avenue.

Poquessing Creek Drive

See Torresdale Avenue.

Porcelain Street

See Moravian Street.

Porter Street

A Pennsylvania governor, David Rittenhouse Porter (1788–1867), who, despite major accomplishments of his administration, was almost impeached, follows his predecessors in the line of governors down Broad Street (see Governors).

Porter pursued the study of law as a young man; at the same time, he bred horses and cattle, and was a partner in the Sligo Iron Works in Huntingdon County. When the Iron Works failed, Porter looked for other sources for income.

Serving in the state legislature from 1819 to 1823, the young man developed a reputation that helped elect him to the state senate in 1836—and to the chief executive's post in 1839. During his two terms in office, Porter suppressed the anti-Catholic riots, upheld the state's credit, and stirred up the antagonism of the legislature. His political enemies tried unsuccessfully to impeach him in 1842.

With his close friend General Sam Houston, he tried to construct a railroad through Texas to the Pacific coast. The outbreak of the Civil War stymied this attempt.

Porter Street appears on city records, from Dover to 28th Street, as being deeded to the city in 1884. The street grew to its present proportions from then until the early twentieth century.

Port Royal Avenue

Port Royal Avenue remains as a link with the colonial past, when major trading was conducted between Philadelphia and the West Indies. A colonial mansion bearing the same name was, for many years, the home of the Stiles and Lukens families.

Running from Umbria to Wissahickon Avenue, Port Royal Avenue was first confirmed in 1772 from Hagys Mill Road to Ridge Avenue. The section from Ridge to Township Line Road was confirmed in 1882.

PORT ROYAL AVENUE. The first master of "Port Royal House," Edward Stiles (donated his name to Stiles Street) was involved in shipping between Bermuda, where he was born, and the colonies. He bought the plantation near Frankford before the Revolution and named it after his birthplace. He lived there in summer, "surrounded by his slaves, assuming the state becoming a great shipping merchant. . . ." Stiles' slaves were freed at the time of his death, as stipulated in his will. They were also granted free education at the estate's expense.

Potter Street

Potter Street, from A Street to Cayuga, was named after the Pennsylvania county (see Counties). Deeded to the city between 1884 and 1957, Potter added only one section of street (A to Huntingdon Street) by name change. Knorr Street became a part of Potter in 1895.

Potter County was formed in 1804 from Lycoming County territory. The county was named for General James Potter (1729–89), who served with distinction during the Revolution.

Powell Street

See Delancey Street.

Powelton Avenue

Samuel Powel (1739-93), the last mayor of Philadelphia before the Revolution and the first after it, left a large estate—Powelton—on the west side of the Schuylkill at Market Street. It was from Powel's estate that the street obtained its name.

Powel's nephew and heir, John Hare Powel (1786-1856), played a major role in the formation of the Pennsylvania Agricultural Society. During the nephew's ownership of Powelton, the first sections of this avenue, from 40th to 42nd Street, were deeded to the city. By 1877, Powelton Avenue was complete to its modern distance.

Preble Street

Honors Edward Preble (1761-1807), a commodore in the U.S. Navy who engaged at Tripoli; he also taught Bainbridge and Decatur.

Price Street

A Philadelphia lawyer who saw the need for consolidation of the legally distinct districts into the City of Philadelphia, and acted upon it, lent his name to this street.

Eli Kirk Price (1797-1884), a descendent of a Welsh associate of William Penn, was the author of the "Consolidation Act" of 1854, which is believed to be the foundation for the growth, development, and importance of the City of Philadelphia (see Consolidation). His civic efforts also led to the establishment of Fairmount Park in 1867. Price served as its first chairman. His descendents, to the present century, have been represented on the board of the Fairmount Park Commission.

Price Street

"Few if any American lawyers devoted so much of their time and experience," Price's obituary read, "to the improvement of the law." He was considered an expert on real estate law and spearheaded the drive to protect the rights of wives and children in abandonment and divorce cases.

Lehman Street, which existed as early as 1867, was changed to Price in 1895. Price Street, as a street name, has been on the official street records only since 1890, two years after his death. According to other sources, the road that was opened through Price's Germantown property in the mid-eighteenth century was the basis for the naming. This is a distinct possibility, since affidavits have been found which indicate that the street in that neighborhood was in use from 1864 or before.

PRICE STREET. Eli Kirk Price, a Philadelphia lawyer with vision, was the architect of Philadelphia's Consolidation in 1854. (Scharf and Westcott's *History of Philadelphia, 1609–1884*)

Priestly Street

PRIESTLY STREET. The Philadelphia misspeller strikes again! This street was named for the "discoverer of oxygen," Dr. Joseph Priestley. What happened to the "e" in his name remains a mystery.

This street was named in honor of Joseph Priestley (1733–1804), chemist and clergyman.

Priestley's scientific research resulted in the discovery of ammonia, sulphur dioxide, hydrogen chloride, and oxygen. To some, his most important contribution was the invention of carbonated water—without which we would not have Coke, Pepsi, Dr. Pepper, and all the rest. Because of his strong views on the French Revolution, he was forced to flee to the United States. Priestley was also the founder of the Unitarian Church.

Somewhere along the way, the "e" in Priestley's name was dropped.

Prime Street

See Washington Avenue.

Prince Street

See Girard Avenue.

Pulaski Avenue

Casimir (Kazimierz) Pulaski (1747–79), for whom this avenue is named, was well known in Europe as a military hero, based on his defense of Czestochowa against the Russians in 1770–71. When the Russians were joined by the Austrians and Prussians, Pulaski realized that resistance was futile and went into exile.

After four years of inactivity, he was introduced to Silas Deane and Benjamin Franklin, who arranged for Pulaski to join Washington. He fought at Brandywine and Germantown before resigning his command over a dispute with Anthony Wayne. Allowed to create the Pulaski Legion, his own unit of cavalry and light infantry, he once again got into petty disputes with other officers.

Pulaski Avenue

Pulaski might have been forgotten by the history books—and the street namers—if not for his death at the siege of Savannah. Pulaski, in a heroic pose, charged the British lines at the head of his cavalry, only to be cut down. He died of his wounds two days later.

Quarry Street

Neither research nor the study of early Philadelphia maps has uncovered the existence of a quarry of any type in the area of this street, thereby discrediting a popular notion. So the most likely source for the name is Judge Robert Quary, who probably owned the section of the street from Bread to 3rd Street. The area which Quary apparently owned is the earliest recorded section to bear the name and was indicated by the city as being "laid out by original owners."

In 1895, Quarry Street grew by the addition of such other named streets as Drinker's Alley (Front to 2nd), Cresson (5th to Sheridan, vacated by 1966), McAllister and Mint Court (8th to a dead end west of 9th, vacated in 1968), Wheat and Rye (Clifton to a dead end west of 10th), Mark's Lane (11th to 12th), Shelbark (13th to Juniper), Landreth (12th to 13th), Green's Court and Agree Alley (Mole to 16th, vacated in 1973), and Toland (20th to a dead end east, vacated in 1955).

As head of the Admiralty Court, Judge Quary was the focus of the controversy over the imprisonment of pirates Kidd, Blackbeard, and Avery. Though ostensibly prisoners, they were able to come and go as they liked.

Queen Lane

QUEEN LANE. This roadway was officially recognized in 1773. The name is derived from the Indian Queen Tavern that once stood at the corner of Germantown Avenue and today's Queen Lane. (Painting by E. T. Scowcroft)

As early as 1692, Queen Lane existed—as the "Cross Street to Schuylkill." At various times during its existence, it has also been known as Bowman's Lane, Riter's Lane, Indian Queen Lane, Fall's Lane, Palmer's Lane, and Whittell's Lane.

Though "the various courses and distances [of Queen Lane] were definitely fixed" in 1723, the lane was not made a matter of record until 1773, when the roadway was confirmed from Germantown Avenue to Ridge. Later, other sections were added, including Conrad to Cresson (1874), Stokley to Conrad (1889), Rufe to Wakefield (1909), and Clarkson to Magnolia (1924). In order to eliminate confusion between this street and Queen Street in Southwark, this Queen became a lane—between Germantown and Wissahickon Avenues in 1917, and between Wissahickon and Conrad in 1925.

The source of the name is the Indian Queen tavern that once stood at the corner of this street and Germantown Avenue. Some sources imply the lane commemorates Queen Anne of England.

Queen Street

In 1653, Lieutenant Swen Shute (later Swanson), who gave his name to Swanson Street, received about eight hundred acres of land, embracing Kingsessing, Passyunk, and Wiccaco in Southwark, from Queen Christina of Sweden (see Christian Street). It was in this queen's honor that the street was named. Folk historians have long held that the street commemorates Catharine the Great because of the proximity of Queen and Catharine Streets. But it seems likely that the Swedish influence on the District of Southwark would have been felt more than that of the Russians.

Queen Street, from Delaware Avenue to 2nd Street, was opened before 1787. There is a distinct possibility that the street might have been laid out and in use earlier than 1767, when Swanson Street was confirmed. The remainder of Queen Street, from 2nd to 6th Street, was confirmed in 1806. The line of the street has remained unchanged since that time.

R

Race Road

See Holmesburg Avenue.

Race Street

One of Penn's original streets (see Original Streets), Race Street was officially known as Sassafras until the middle of the nineteenth century. When Thomas Holme arrived in Philadelphia, he named this street Songhurst for John Songhurst, a prominent Quaker preacher who arrived here in 1682. On his second visit to his colony, Penn struck out Songhurst—and many other street names which commemorated individuals—and dubbed the street Sassafras.

Though many historians consider all the original streets of the "checker-board" as having been laid out at the same time, the latest information indicates that the building lots on the south side of Sassafras were laid out to 4th Street from 1704 to 1707, from 4th to 5th between 1686 and 1717, and from 5th to 6th in 1714. Though the street may have existed on the plan of 1682, it seems highly unlikely that a physical street existed prior to 1686; more probably, it came into being in the eighteenth century.

During the eighteenth century, the young men of the city used Sassafras as a race track. The roadway, which also led to the race course around Centre Square (now City Hall), became increasingly dangerous until 1726. The Grand Jury of that year reported that "since the city has become so very populous the usual custom of horse racing at fairs in the Sassafras st. is very dangerous to life."

Philadelphia legislated to change the official name from Sassafras to Race in 1853. After the Consolidation of 1854 (see Consolidation), the West Philadelphia continuation of the street, then known as Huntingdon, was also changed to Race.

Randolph Street

Randolph Street was named for Jacob Randolph (1796–

1848), a surgeon, born in Philadelphia. He was married to Sarah Emlen Physick, daughter of the famous Dr. Philip Syng Physick.

Reckless Street

See League Street.

Redfield Street

See Alden Street.

Reed Street

Reed Street came into being on January 5, 1790, by order of the commissioners of the District of Southwark. In their extensive brief of that date, the commissioners neglected to indicate for whom the street was being named.

The most logical person for the honor is Joseph Reed (1741–85), who during the early days (1775–76) of the War for Independence served as aide-de-camp and military secretary to General Washington. He resigned that post to become a member of the Continental Congress (1777–78). Later, he was elected president of the Supreme Executive Council of Pennsylvania. As president of the council, he led the state's attack on Benedict Arnold and played a key role in settling the "Mutiny of the Pennsylvania Line." Though he served in key patriotic roles, his loyalty to the 1776 cause has been disputed. Reed has been called, by his contemporaries, a "trimmer," one who changes his political loyalty to suit his personal interests.

Reed Street grew from its initial eighteenth-century span—from 3rd Street to the Delaware River—to its present length mostly in the late nineteenth century. The last official opening of the street is dated June 5, 1913.

Rex Street

See Summit Street.

Richard Allen Way

Richard Allen (1760–1831) was a Philadelphia-born slave who, at the age of 22, was licensed to preach in the Methodist Episcopal Church. Four years later, Allen had saved enough money to purchase his freedom. He joined St. George's Methodist Church, but was only permitted to preach to the black members.

Allen apparently mesmerized his congregation, because its number swelled to the point where the white congregation became angry. Allen and his supporters left St. George's, converted an old blacksmith shop into a church, and organized the Free African Society, the first black church in America.

In 1794, the congregation built a new church and became part of the Methodist Episcopal Church. The church was dedicated by Bishop Francis Asbury. Five years later, Richard Allen became the first black to be regularly ordained in the Methodist Church. He was also a driving force in the development of the African Methodist Episcopal Church and, in 1816, became its first bishop.

The City of Philadelphia named a portion of South 6th Street leading to Mother Bethel A.M.E. Zion Church, the mother church of Allen's denomination, in his honor. "Richard Allen Way" sits atop the 6th Street signs, and residents still use 6th Street as their address.

Ridge Avenue

Like so many of the early roads leading out of Philadelphia, Ridge Avenue followed an old Indian trail. It was so named because it was situated on the ridge between the Schuylkill and the Wissahickon. The first settlers referred to it as the Manatawny or Plymouth Road, because it led toward the Manatawny Creek and to Plymouth Meeting.

In 1803, the citizens of the area petitioned the legislature for a turnpike road (see Toll Roads) along the ridge. This petition was refused because the Germantown Turnpike ran parallel to it. Eight years later, an act was passed "to enable the government to incorporate a company for making an artificial road beginning at the intersection of Vine and Tenth Street" and running on to the Perkiomen. The route was to be "as near as may be consistent with economy and utility, to Wissahiccon creek; thence to Barren Hill; thence to Norristown in the County of Montgomery; and thence by the nearest and best route to the Perkioming bridge in the county aforesaid."

The Ridge Avenue Turnpike Company must have acted swiftly because the Ridge Road is listed in the street directories beginning in 1813. The road was extended by the jury in 1836 to cover the span from Vine to Cambridge. The remainder was opened by affidavit between 1883 and 1904. The turnpike was freed from toll prior to 1873.

Ridge Street

See Gilligham Street.

Rising Sun Avenue

Rising Sun Avenue takes its name from an early village of Philadelphia located at the intersection of Germantown Avenue, Old York Road, and Westmoreland Street. The village, originally called Sunville, was settled by two young men, Heinrich Frey and Joseph Plattenbach, in 1680. At that time, they operated a blacksmith shop at the present corner of Front and Arch Streets. A frequent visitor, tradition tells us, was Joseph, son of Chief Tamane of the Lenni-Lenape. At Joseph's suggestion, the two young smiths followed him through the woods to the tribe's headquarters. There they were made adopted tribal members. Before they returned to the city, Tamane is supposed to have taken them to a spot near where Germantown Avenue meets Old York Road and told them the tribal council had decided that all the land in that region was theirs—"until the Great Spirit should call them to the Eternal Wilderness." At that moment, the tale continues, the sun was rising in the east and the young men named the spot Aufgehende Taune or Rising Sun.

Another source for the name is the old Rising Sun Inn or Tavern, which was opened in 1754, burned by the British in 1777, and demolished about 1888. The inn was established, however, long after Frey and Plattenbach settled in the area.

Rising Sun Avenue was first confirmed by the road jury of Quarter Sessions Court in 1789 from Old York Road to Fisher's Lane (present-day Ruscomb Street). It was extended in 1814 as far as Cottman Avenue. During its early days, Rising Sun Avenue was a toll road known as the Kensington and Oxford Turnpike (see Toll Roads). The road was freed from tolls in the early twentieth century. The remaining sections of the avenue were put into use after it ceased being a toll road. Segments were deeded to the city between 1919 and 1965.

Riter's Lane

See Queen Lane.

Ritner Street

The thrifty son of a German immigrant, who hauled his own produce and drove his own stock to market in Philadelphia, and later went on to become governor of Pennsylvania (see Governors), is commemorated by this street.

Joseph Ritner (1780–1869) virtually pulled himself up by the bootstraps. The son of a German settler and ardent Revolutionary patriot, Ritner's formal education consisted of six months' schooling and instruction in weaving. Despite that, he served in the State Assembly from 1821 to 1826, the last two years as speaker. He ran for governor four times, winning only a single term. His administration, by today's standards, could be considered lackluster.

Rittenhouse Street

The street which bears his name is mainly the result of acquisitions during the late nineteenth century.

Rittenhouse Square

To a native Philadelphian, a Rittenhouse Square address is tantamount to social acceptance. But the elegance we attribute to this one of the only two Philadelphia streets to be called squares has not always existed.

The park area of Rittenhouse Square is one of the original five (see Squares) designated for public use by William Penn. By the early eighteenth century, a law setting a five-shilling fine for shooting fowl on the streets of Philadelphia chased hunters to "The Governor's Woods," the present-day square.

Dirt from city streets was dumped in the woods for years, and pigs, chickens, and cows grazed through it. Brickmakers dug for clay and the resultant quarry holes turned into ponds. The ponds drew duck and geese. The area, as a result, was nicknamed Goosetown.

In 1825, the public square was renamed in honor of David Rittenhouse, a Philadelphian who became America's first astronomer. To reflect the square's new designation, an 1867 ordinance changed the names of two small streets (Plymouth, from the square to 20th Street, and Murray, from 20th to 21st) to Rittenhouse. Three decades later, in an effort to unify street nomenclature, Rittenhouse Street became Irving. Finally, Irving Street, the bed of which had been laid out and dedicated to the city in 1834, became Rittenhouse Square in 1897, at the height of the square's social status.

RITTENHOUSE SQUARE. "The Governor's Woods," as it was first known, became Rittenhouse Square in 1825 to honor David Rittenhouse, a Philadelphian who became America's first astronomer. David is not to be confused with William, the first bishop of the Mennonite Church, for whom Rittenhouse Street in Germantown was named.

Rittenhouse Street

William Rittenhousen (later Rittenhouse), the first bishop of the Mennonite Church, immigrated to Germantown in 1686. Four years later, he built the first paper mill in America on a little stream, the Monodhone Creek (later called Paper Mill Run), which flowed into the Wissahickon.

Rittenhouse Street

A road to the paper mill must have existed prior to 1760, the earliest confirmation on record, from Wayne Avenue to Greene Street. An act of assembly in 1871 opened the way from the Wissahickon to Wayne.

In 1895, Centre (Germantown Avenue to Magnolia) and Ladley (McMahon to Chew Avenue) were changed to Rittenhouse. West Rittenhouse Street had another name, Poor House Lane, which, unfortunately, city records do not reflect.

Three centuries later, William Rittenhouse's original stone house still stands on Wissahickon Drive at Rittenhouse Street. Rittenhouse's grandnephew was David Rittenhouse (see Rittenhouse Square), America's first astronomer.

RITTENHOUSE STREET. William Rittenhousen (later Rittenhouse) was the first bishop of the Mennonite Church, and built the first paper mill in America. His original stone house is still standing on Wissahickon Drive at Rittenhouse Street. (Philadelphia Municipal Archives)

River Road

See Schuylkill Avenue.

Robin Hood Dell

See Ford Road.

Rock Street

See "Things that Grow . . ."

Rockland Street

See Brandywine Street.

Roosevelt Boulevard

The Boulevard honors Theodore Roosevelt (1859–1919), the twenty-sixth president of the United States.

Rosemary Lane

Rosemary Campbell's father, John T., was a district surveyor for the streets department. In 1930, Campbell arranged to have the name of Sedgwick Street, now running from Emlen Street to Lincoln Drive, changed to honor his daughter.

Roxborough Avenue

Originally called Conrad's or Collins's Lane, Roxborough Avenue was first confirmed in 1761, from Ridge Avenue to Cresheim. That confirmation covered the present-day roadway from Ridge Avenue to Magdalena.

The avenue was named after the section of Philadelphia which bears the same name. The section was formerly known as Rocksborrow, supposedly because of the burrows among the rocks made by foxes.

In the section from Terrace to Mitchell, the street—then known as Mechanic—was deeded to the city in 1871. In the same year, council resolved to change the name to Roxborough between Terrace and Bellair. Twenty-four years later, Mechanic Street from Main to Boone was also changed.

With a grade of 17 percent, Roxborough Avenue is

the steepest street within the limits of the City of Philadelphia.

Roxborough Street

See Lycoming Street.

Rubicam Street

Allen Rubican, a carpenter, lived at James and Rubican's Place (near 927 Noble Street) in 1850. As the Rubican family grew and moved into the Olney and Germantown sections of Philadelphia, the streets named in their honor took on the altered spelling which the family had assumed by 1872.*

The earliest recorded section of today's Rubicam Street is from Stenton to Wister, deeded to the city in 1872. The section from Wister to Bringhurst, confirmed in 1875 as Mercer Street, became Rubicam in 1895. The final segment, the block from 2nd to 3rd Street, was deeded in 1914.

Ruffner Street

Named for the Ruffner family, owners of land in the Nicetown section of Philadelphia in the nineteenth century, this street came into existence in 1895 by the change of the names of Midvale (Schuyler to Blabon Street) and Howard (Germantown Avenue to 16th Street, vacated in 1970). From 1897 to 1941, Ruffner Street increased to its present length.

The records of St. Stephen's Roman Catholic Church, established in 1843 to serve the Nicetown area, indicate that the first child to be baptized there was William Anthony, son of William Anthony and Elizabeth Ruffner. The church then stood at Barr and what is now Lycoming Streets. It has since moved from that area.

Rubican is French for a mixture of black and white hair.

Rundy (Rundle) Street

See Delancey Street.

Rye Street

See Quarry Street.

Ryers Avenue

See Burholme Avenue.

Sage Street

See Jefferson Street.

St. Davids Street

See Bonsall Street.

St. Georges Road

To determine for whom or what a street was named can sometimes bring the researcher to the brink of despair or insanity. The search for the "St. George" of St. Georges Road is one example.

Gowen, from Cresheim Valley Drive to McCallum Street, was changed to St. Georges in 1925, only nineteen years after the street was opened. Research of maps of the area show no church by that name close by, and no inn. Nor was there any geographic caricature of a dragon.

Then, by happenstance, an index card was located in the streets department which read:

From Mr. Campbell:
St. George's Road named by Morris Llewellyn Cooke, presumably in honor of George Woodward, M.D. Cooke wanted an English name and also wanted to honor Dr. Woodward.

St. James Street

See Willings Alley.

St. John Street

See American Street.

St. John Neumann Place

John Nepomucene Neumann (1811–60) was the fourth Roman Catholic bishop of Philadelphia. During his tenure, he expanded church membership, created the parochial school system, and defended his church's rights during the Anti-Catholic Riots of the 1840s in Philadelphia. Philadelphia's Cathedral of St. Peter and Paul shows signs of Bishop Neumann's activities. The cathedral was being built during the riot period, and Neumann wisely ordered the workmen to brick up the windows on the ground level as protection against rock throwing. The windows were never unbricked.

Neumann also helped establish the Sisters of Notre Dame and the Sisters of the Third Order of St. Francis. He was canonized by the Roman Catholic Church in 1977.

St. John Neumann Way

See St. John Neumann Place.

St. Paul Street

See Fulton Street.

Salmon Street

See "Things That Grow . . ."

Sandpiper Place

See "Things That Grow . . ."

Sansom Street

In the midst of Isaac Norris's pasture (Chestnut and Walnut, between 8th and 9th), Robert Morris began construction of a marble mansion. Six years—and debts exceeding $75,000—later, Morris saw his property and his incomplete mansion sold at sheriff's auction. Most of the property was sold to William Sansom, to whom Morris also owed sizable sums of money.

With the aid of architect Benjamin Henry Latrobe, Sansom designed a totally new concept in home building, the Philadelphia row house. He built two rows of houses back to back in 1803 between 7th and 8th. There were twenty on Walnut and twenty-two on the south side of present-day Sansom, which he dubbed Sansom's Row. Unfortunately, contemporaries thought that Sansom's houses were in an area "too remote and lonely." At the time they were right; Walnut Street was not paved west of 6th Street. In an effort to rent or sell his properties, Sansom paved his own street. He even offered to pay the city to extend Walnut the additional two blocks.

In 1849, Little George Street, which ran from 6th to 7th, became Sansom Street. Nine years later, George, from 8th to 24th Street, followed suit. Gothic and Lodge (Front to American), which also predated Sansom, were changed in 1895 to commemorate the builder. Library, from 4th to 5th, was changed to Sansom that same year, but the ordinance was amended to keep it Library.

York Street in West Philadelphia became Sansom Street after the 1854 consolidation (see Consolidation). As far west as 46th Street, Sansom existed in use before the turn of the twentieth century. The remainder of the street, to Cobbs Creek Parkway, was added before 1924.

Sartain Street

One of America's foremost mezzotint engravers brought worldwide attention to his adopted Philadelphia and, for his efforts, had a street named after him.

John Sartain (1808–97) came to America from England in 1830. Shortly after his arrival, he produced the first important mezzotint in the United States. Several unsuccessful publishing ventures forced him into creating vignettes and engravings for early bank notes. Three years after he attended the 1855 International Exposition in Paris, the name of Margaretta Street, from Poplar to Girard, was changed to honor him.

In 1875, Sartain was selected chief of the bureau of art for the Centennial, a position which drew to him universal acclaim. The designs for the monuments to Washington and Lafayette at the old Monument Cemetery were products of his genius.

Two years before his death, City Council changed the names of several streets in line with Sartain. These were Gerhard, from Ritner to Moyamensing and from McKean to Moore; McIllery, from Hall to Montrose; Beckwith, from Catharine to Bainbridge; Manship, from Manning to Locust; Sussex, from Spring to Vine; Marvine, from Stiles to Thompson; and Hibberd, from Girard to Stiles. All these early streets dated back to the mid-nineteenth century. The section from Oregon Avenue to Shunk Street was added in 1914.

SARTAIN STREET. John Sartain, one of America's foremost mezzotint engravers, received a rare honor in 1858 when the City of Philadelphia renamed Margaretta Street in his honor. Few people live to see their names on streets.

Sassafras Street

See Race Street.

Saxton Street

See Wishart Street.

School House Lane

Originally an Indian trail which led from the Wingohocking Creek to Ridge Road, at the mouth of the Wissahickon, School House Lane was used by the people of Germantown "with great conveniency . . . for upwards of 30 years last past [before 1693] but only on sufferance." It was frequently referred to as the "Cross Street," or the "Cross Street to Schuylkill."

The roadway, then called Bensell's (or Bensil) Lane, was laid out and confirmed by the Quarter Sessions

School House Lane

Court in 1760. The timing of the road approval was excellent. Only two years before, a meeting had been held in Germantown at which it was resolved to erect a building for an "English and High Dutch or German School." It was opened in 1761 as the Germantown Union School-House, built on ground purchased from John and George Bringhurst (see Bringhurst Street) for £125.6. The school continues today as the Germantown Academy, though not at the same address. The school's bell, incidentally, came to this country aboard the *Beaver*, the same vessel that carried the party favors to the famous "Boston Tea Party."

After the school was erected, Bensell's Lane became known as School Lane. In 1893, School Lane became officially School House Lane.

SCHOOL HOUSE LANE. The Union School House of Germantown was founded in 1759. The road that led to the school took on the directional name. The Union School House became Germantown Academy. Germantown Academy no longer resides in Germantown, though the street name remains. (Scharf and Wescott's *History of Philadelphia, 1609–1884*)

School Lane

See School House Lane.

School Street

See Parrish Street.

Schuyler Street

Schuyler Street was named in honor of Philip John Schuyler (1733–1804). Schuyler, a native of Albany, New York, was one of four major generals to serve under Washington at the start of the Revolutionary War. His military exploits did not enhance his reputation. He was in command of the failed 1775 invasion of Canada and, in 1777, was blamed for Arthur St. Clair's surrender of Fort Ticonderoga. He was later exonerated for his actions in the loss of Ticonderoga. Following the war, he served in New York's senate and, following the ratification of the federal Constitution, he became one of New York's first U. S. senators.

Schuylkill Avenue

Taking its name from the river at its side, Schuylkill Avenue dates back to the early nineteenth century, when it was called Point Breeze or River Road. Another old section, from present-day Pennypacker Avenue to Hartranft Street, was opened in 1810 and called Gallow's Lane. This old road ran from Beggartown Lane to the Penrose ferry. The name Schuylkill, which means "hidden stream," was given by early Dutch explorers who passed the mouth of the river without seeing the stream.

The main extension of Schuylkill Avenue took place in 1860 with land deeded to the city by Andrew W. Eastwick (see Eastwick Avenue). Eastwick's land for this road extended from today's Morris Street to Grays Ferry Avenue. The northern reaches of the avenue, from Ellsworth to 27th Street, were confirmed in 1834, south to the U.S. Arsenal, on the avenue then known as Sutherland.

Sedgewick Street

See Rosemary Lane.

Seminole Avenue

See Indian Tribes.

Seneca Street

See Parrish Street.

Seybert Street

See Ingersoll Street.

Shackamaxon Street

This street not only preserves an old Indian name, but also leads to a famous Kensington town or neighborhood where the sachems or chiefs gathered. Shackamaxon is older than Philadelphia itself. Laurens Cock, at a Swedish court held November 12, 1678, acknowledged the conveyance of three hundred acres of land lying "on the west side of the Delaware River, at the towne or neighborhood called and known by the name of Sachamexin—the whole dividend or quantity of land being of late surveyed for the inhabitants of Sachamexin in general, and containing 1800 acres." William Penn met with the Indian chiefs at Shackamaxon, now known as Penn Treaty Park, where his famous treaty was ratified. To set the record straight, we should note that Penn never signed the treaty with the sachems because, as he put it, "A treaty never signed is never broken."

The English corruption of Sachamexin has led to a wrong translation of the word. Indian scholars, using the original word, translate it to mean "meeting place of chiefs." Others, using the corrupted word, come up with "place of eels."

The street itself was opened by the road jury in 1816, from Richmond to the Delaware River. Shackamaxon Street was widened in 1838. The entire existing stretch, from Frankford Avenue to Penn Street, was in public use before 1862.

SHACKAMAXON STREET. Benjamin West's famous painting, "William Penn's Treaty with the Indians" at the place known as Sachamexin (Shackamaxon), is a fraud! Penn never signed a treaty with the Indians. In fact, as Penn put it, "A treaty never signed is never broken."

Shady Lane

See Grant Avenue.

Sharpnack Street

See Upsal Street.

Shawnee Street

See Indian Tribes.

Shelbark Street

See Quarry Street.

Shinbone Alley

See Manheim Street.

Shippen's Lane

See Bainbridge Street, Manheim Street.

Shippen's Street

See Bainbridge Street.

Shisler Street

See Agusta Street.

Shunk Street

See Governors.

Skidoo Street

Leave it to Philadelphia street namers. In 1982, Councilwoman Ann Land introduced a bill that changed the name of a one-block street in East Falls from Cresson Street to Skidoo. East Falls residents have always called it "Skidoo." No one has any idea why! Land's bill only made it all official.

Snyder Avenue

See Counties, Governors.

Somerset Street

See Counties.

Somerton Avenue

Somerton Avenue was named after the section of Philadelphia through which it runs. A product of the early twentieth century, Somerton originally ran from Bustleton Avenue to Audubon. Today, it extends only from Bustleton to De Pue.

Songhurst Street

See Race Street.

South Street

Originally designated by Penn as Cedar Street (see Original Streets), this street was commonly known as South as early as 1707 because it formed the southern boundary of the City of Philadelphia, separating it from the District of Southwark. It was not until 1853 that the popular name became official.

Extending from Delaware Avenue to 33rd Street, South Street, though one of Penn's original streets, was not legally opened until 1883.

Spring Garden Street

Spring Garden Street was named for the district through which it ran. The District of Spring Garden (see Consolidation), originally bounded by Vine Street, Fairmount Avenue, Broad, and 6th Street, probably derived its name from Spring Garden, the country seat of Dr. Francis Gandouet, which was located near 7th and Buttonwood as early as 1723.

Though records indicate that Spring Garden Street was first confirmed in 1833 from 6th to Broad, permission was given in 1811 "to open a road to the west side [of the Schuylkill] from Lancaster Pike to the bridge, which is the present Spring Garden Street." This indicates a much earlier—if not legally opened—roadway. The road jury confirmed the segment from Lancaster to Callowhill Bridge in 1834.

By the end of the nineteenth century, Spring Garden Street extended as far east as 6th Street, including the sections in the possession of the owners of the Bush Hill Estate (16th to 18th Streets). From the Delaware River west, the street was opened by ordinance in 1908. This stretch of road contains portions of Green Street, confirmed by jury in 1785.

There is an interesting footnote regarding this street. In 1858, the name of Morris Street, from Broad to Fairmount, was changed to Spring Garden. Surprisingly, Spring Garden Street does not extend as far as Fairmount. It stops at East River Drive.

Spruce Street

See Original Streets.

Squares

"In the centre of the city is a square of ten acres; at each angle are to be houses for public affairs, as a meeting place, Assembly or State House, market house, schoolhouse, and several other buildings for public concerns. There are also in each quarter of the city a square of eight acres, to be for the like uses, as Moorefields in London." That was Penn's concept for setting aside green spaces in his 1683 town. Noble it was. But his use of the present tense, implying that the squares were actually there, was putting a varnish on the truth. The five squares did not exist—except on paper—at that time. They were, however, preserved from commercial or residential use.

The names for these squares did not come into existence until 1825. Penn Square, named for William Penn, was called Centre Square because it was in the center of the proposed town. Used as a military drill field and later as the site for the municipal water works, it is now occupied by Philadelphia's City Hall. Logan Square was named after James Logan, "one of the early settlers of the state, secretary to William Penn and founder of the Loganian Library in this city." The main branch of Philadelphia's Free Library is located there. Rittenhouse Square, used during the late eighteenth and early nineteenth century as a depository for dirt and debris, and later as a source for brick clay, was named for David Rittenhouse, "the enlightened philosopher" (see Rittenhouse Square). Washington Square, the location of the Tomb of the Unknown Soldier of the Revolution and the site of Philadelphia's first potter's field, honors George Washington. Franklin Square, named for philosopher and stateman Benjamin Franklin, has long been a resting place for derelicts and residents of the city's skid row. It is the first public square seen by travelers coming to Philadelphia from New Jersey.

The final city square, not listed by Penn in his prospective, is Independence Square, the most hallowed spot in Philadelphia—the site of the old State House, where the Declaration of Independence was born and where the bell first "proclaimed liberty throughout the land."

Other squares have popped up in Philadelphia's growing plan to add a touch of green to the red of the brick and gray of the granite buildings which have changed Philadelphia from a "green countrie town" to a city of homes.

State Road

See Tacony Street.

State Street

See Tacony Street.

Stein Street

See Ingersoll Street.

Stella Maris

See Pollock Street.

Stenton Avenue

See Wingohocking Street.

STENTON AVENUE. James Logan's mansion was called "Stenton." The garden porch is shown in this 1922 photograph. It was in this garden that Chief Wingohocking tried to exchange names with Logan—an Indian token of friendship. Logan suggested they name the stream that ran through his property instead. The chief bought it, and Logan never became James Wingohocking.

Stirling Street

This street is named for William Alexander (1726–83), the self-styled "Lord Stirling" who served under Washington during the Revolution.

General Alexander had a military career of distinction, but is better remembered for reporting the Conway Cabal—an attempt to replace Washington with Horatio Gates—to Washington, his presiding over the court-martial of Charles Lee for his performance at Monmouth Court House, and his role in the board of inquiry on Major John Andre.

Stockton Street

See Iseminger Street.

Stokley Street

Stokley Street, from Westmoreland to Coulter Street, was named for William S. Stokley, three-term mayor of Philadelphia. Stokley (1823–1902) was instrumental in the reorganization of the Philadelphia police and fire departments during his tenure in office. His son, Horatio N., was a real estate assessor at the time of his father's death. It is possible the son assisted in getting this street named.

The first deed for the roadbed between Queen Lane and Coulter and between Coulter and School House Lane (vacated in 1930) was conveyed to the city in 1907–8. The remaining sections were added before 1920.

Stone House Lane

Stone House Lane, of which only a fragment of the original remains, takes its name from a single stone house erected there during the eighteenth century. According to tradition, the lane was settled by Hessian mercenaries who remained in America following Cornwallis's surrender at Yorktown in 1781.

Stone House Lane

Not shown on any directory until 1906, Stone House Lane was a haven for squatters, who built tar-paper shacks throughout the "meadows" of southwest Philadelphia's Eastwick section. It took the City of Philadelphia nearly thirty years to rid the area of these interlopers. The last was removed in 1956.

Stone House or Ark Lane was opened in 1763 from old Terminal Avenue to Porter Street. Vacated between 1926 and 1958, the lane remains as indentations in the roads which followed it.

Strawberry Street

Strawberry Alley, surveyed by Penn's planners from 1690 to 1692, is officially known as a street, even though it is only a block long and barely wide enough to drive an auto through.

The street is undoubtedly named for the strawberry plant. Tradition has it that an early settler or two planted these delicacies along the way. It is more feasible to assume that wild strawberries were growing there when the settlers arrived, and the name would thus fulfill Penn's desire to name the streets after things which grew wild in the area.

In 1895, the city changed the name of the alley, from Market to Chestnut, to Philip Street. Two years later, it resumed its original name.

Strongford Street

See American Street.

Stump Street

See Cemetery Avenue.

Suffolk Street

See League Street.

Summit Street

Named in 1893, Summit Street is the highest point in Philadelphia County from Trenton to Bryn Mawr—446 feet above sea level. Prior to being known as Summit, the street was called Rex.

Susquehanna Avenue

See Counties.

Sussex Street

See Sartain Street.

Sutherland Avenue

See Schuylkill Avenue.

Swanson Street

See Queen Street.

Sylvan Street

See Wyalusing Avenue.

T Tacony Road

See Torresdale Avenue.

Tacony Street

Named for the Tekony or Tekone Creek, which, in the tongue of the Delaware Indians, meant "forest or wilderness," Tacony Street was first confirmed by the Court of Quarter Sessions in 1703, from the Tacony Creek to Frankford Mills.

The street is one of the oldest in the city and appears on maps as early as 1700. Added to and improved over the years. Tacony is now the approach to the Tacony-Palmyra Bridge. That final section, then called State Road, from Lardner to the bridge, was confirmed in 1894 and the name was changed to Tacony. The street was also known as State or Aramingo from the Little Tacony Creek to the railroad, and as River Road from the Frankford Creek to the Pennypack Circle.

For some time in the 1880s, Tacony was the terminus of the Trenton railroad. Travelers took a steamboat from there to the Walnut Street wharf in downtown Philadelphia.

Talbot Street

See Delancey Street.

Tasker Street

Thomas T. Tasker (1799–1892), in partnership with Stephen P. Morris (for whom Morris Street was named), erected the Paschal Iron Works at 5th and Franklin Streets in 1836. A civic-minded individual who had served as president of the Citizen's Volunteer Hospital during the Civil War, Tasker was the inventor of the self-regulating hot-water furnace. He was honored by

"Things That Grow... And Are Native"

the city in 1858, when Franklin Street, where his works were established, became Tasker Street.

Franklin Street was opened as "a public highway," from Front Street to the Delaware River, in 1831. The section where the Paschal Iron Works stood was open land until confirmed in 1856. That section of land extended from Front to 7th Street.

The western reaches of Tasker grew during the late nineteenth century, from 10th Street to 26th Street. After the turn of the century, the remainder, from 26th to 33rd Street, was added.

Tatlow Street

See Noble Street.

Taylor Street

See Brill Street.

"Things That Grow... And Are Native"

Those land developers and planners who came after William Penn frequently named new streets in the Penn tradition—after things which grew and were native to the Philadelphia area. There are more than one hundred and fifty such streets in the city representing almost every imaginable species of flora—from Acorn (Street) to Verbena (Avenue). A handful of animals and birds, from Deer (Lane) to Sandpiper (Place), are also included. The fish run from Bass (Street) to Salmon (Street). Finally, there are more than twenty streets bearing the names of precious, and not-so-precious, stones and minerals—from Agate (Street) to plain, old, ordinary Rock (Street).

Thomas Mill Road

Prior to 1793, Thomas Mill Road existed as a road "from Germantown road near Streeper's lane by Woolen's late Dan'l Thomas' now Spruce Mills to the Ridge road at Henry Stroup's." Later, the Quarter Sessions Court confirmed it as a road from Ridge to Park Line.

The road probably began its life in 1737, when Joseph Woolen was permitted to use a road from Roxborough to Chestnut Hill, then known as Spruce Mill Road. Almost half a century later, after the mill had changed hands several times, Daniel Thomas, a miller from Moreland Township, obtained ownership. He and his family owned the property for over fifty years.

Thomas Mill Road once extended from Skye Drive to Clyde Lane. Those sections were vacated in 1967.

Thorp's Lane

See Chew Avenue.

Tiernan Street

See Carlisle Street.

Tilghman Street

Tilghman Street was named for the Tilghman family, whose members include Tench (1744–86), aide-de-camp to Washington who brought the news of Cornwallis's surrender at Yorktown to Congress in Philadelphia, and Richard Albert (1824–99), the Philadelphia-born chemist who developed the use of coal gas in chemical operations.

Tioga Street

As early as 1854, Tioga Street was in use, between Ontario and Venango Streets, in Richmond. The section of the street from Crowell Street to Germantown Avenue was confirmed by the Quarter Sessions Court in 1854. By the end of the nineteenth century, the street had grown to its present proportions.

Applying a popular street-naming custom of the day, the street was named after Tioga County, Pennsylvania (see Counties). Formed in 1804, the county derived its name from the Tioga River, which flows through it into New York. Tioga comes from the Indian *tavego* and *diahoga*, which translate as "the place where the two rivers meet" and "gateway."

Toland Street

See Quarry Street.

Toll Roads

The turnpikes, expressways, and toll roads of the twentieth century are descendents of eighteenth-century vision, or nightmare. The construction of highways has been of great importance from the very founding of this nation. Roadways were needed to carry people from their homes to their place of worship, and to convey goods and services from their place of origin to market. And the need for ways of transporting people and goods was such that travelers and merchants were willing to pay for the use of such thoroughfares.

The first turnpike in America was started in Philadelphia in 1794—from Philadelphia to Lancaster (see Lancaster Avenue). Within the next century, highways sprang up throughout the Commonwealth of Pennsylvania, many originating in Philadelphia. Privately owned, the turnpikes sold stock in the roads and derived their income from the tolls collected from travelers entering and leaving the city. Funerals were permitted free passage over some toll roads, but because of the condition of the highways, many funeral processions were afraid to take advantage of the concession.

Some of the early turnpikes were "plank" roads, constructed of timbers two-and-a-half inches thick laid on wooden stringers two to three feet apart. The road-

Toll Roads

way of wood was elevated above the ground to keep the wood from decaying. It did not provide a very comfortable ride.

A list of the principal toll roads originating in Philadelphia follows.

Original Name	Date Incorporated	Date Freed from Toll	Current Name
Belmont Plank Road			Belmont Avenue
Bustleton and Smithfield Turnpike (Bustleton and Summerton Turnpike)	1804 (1840)	1902	Bustleton Avenue
Byberry and Andalusia Turnpike (Gravel Pike)		1897	Byberry Road
Byberry and Bensalem Turnpike		1900	Haldeman Avenue
Cheltenham and Willow Grove Turnpike			
Chestnut Hill and Springhouse Turnpike	1803	1904	Bethlehem Pike
Frankford and Bristol Turnpike		1892	Frankford Avenue
Germantown and Perkiomen Turnpike (Germantown and Reading Turnpike)	1801	1869	Germantown Avenue
Germantown and Willow Grove Plank Road	1853	1917	Mount Airy Avenue
Kensington and Frankford Plank Road	1857	1873	Kensington Avenue
Kensington and Oxford Turnpike	1814	1899(?)	Rising Sun Avenue
Limekiln Turnpike	1735	1903	Limekiln Turnpike
Oxford Plank Road		1904	Oxford Avenue
Old York Road	1711	1918	Old York Road
Philadelphia, Bala and Bryn Mawr Turnpike		1903	54th Street
Philadelphia, Brandywine and New London Road (Baltimore Pike)	1810	1873	Baltimore Avenue
Philadelphia, Bristol and Morrisville Turnpike	1803		

Philadelphia and Lancaster Turnpike	1792	1903	Lancaster Avenue
Ridge Pike	1812		Ridge Avenue
Tacony Plank Road			Tacony Street

Torresdale Avenue

Before the consolidation of the city in 1854 (see Consolidation), Torresdale was a village in the northeast, near the junction of the Delaware River and the Poquessing Creek, formerly part of Lower Dublin Township. Before the arrival of white settlers, the section was called Poquesink or Poquessing, "the place of mice," by the Indians. That name survives in Poquessing Street and Poquessing Creek Drive.

Charles Macalester (1765–1832) (see Macalester Street) laid out the village on land he purchased from John Risdon, and named it Torrisdale after his family's estate in Scotland. Macalester's mansion was called Glengarry. Later, his descendents sold the land to Samuel Grant (see Grant Avenue).

Torresdale Avenue—and its spelling—is the product of late nineteenth- and early twentieth-century planning. The oldest section of the avenue, from Hunting Park Avenue to Adams Avenue, contains the bed of the old Tacony Road, confirmed in 1703.

Tower Street

See Appletree Street.

Townsend Street

See Fletcher Street.

Township Line Road

See Wissahickon Avenue.

Trenton Avenue

Trenton Avenue by no stretch of the imagination leads to Trenton, New Jersey. Running from Hewson Street to Margaret, its oldest section is from Hewson to Norris. Originally called Maple, then Neff (which never took hold), the street is listed as early as 1854. After Maple became Trenton in 1895, the other sections were added. Trenton Avenue was complete by 1923.

There is no discernible reason for naming the street for the city of Trenton. However, Trenton was founded by William Trent, who arrived in this country in the late seventeenth century. He bought the land where New Jersey's capital stands in 1714 and settled it in 1721. There is a remote possibility that Trent owned land along the route of present-day Trenton Avenue while he was living in Philadelphia in the "slate-roof house" he purchased from Samuel Carpenter. He sold the house to Isaac Norris (see Norris Street) before he began pioneering in the Jersey wilds.

Trinity Place

See Delancey Street.

Tulpehocken Street

Tradition tells us that when the first German settlers arrived in what is now Germantown, they found an Indian footpath on the line of present-day Tulpehocken Street. The name itself is Indian for "the land of Turtles," a clan of the Leni-Lenape tribe. One of the original townships of Lancaster County in 1729, later incorporated into Berks in 1752, bears the same name.

Today's Tulpehocken Street was first recorded in 1850, from Wayne Avenue to Germantown Avenue. Other sections, from Germantown Avenue to Cheltenham Avenue, were added in the twentieth century.

Underhill Street

See McMahon Street.

Union Avenue

See Gravers Lane.

Union Street

See Counties, Delancey Street.

University Avenue

On June 19, 1936, during an official inspection of the area around Convention Hall, now the Civic Center, in preparation for the Democratic National Convention, Mayor S. Davis Wilson lifted the barrier across the newly constructed University Avenue at Vintage Avenue (see Civic Center Boulevard). The official opening took place five years after the ordinance to open had been approved by Philadelphia's City Council. Named for the nearby University of Pennsylvania, University Avenue was widened in 1947.

Upland Way

The only logical reason for the naming of Upland Way is the position of the roadway up above the tracks of the old Pennsylvania Railroad and the Acme Company's warehouse and baking plant. Upland came into being when 60th Street, from Malvern to Overbrook Avenues, underwent a name change in 1908. The section from 54th Street to Malvern was deeded to the city in 1916.

Upsal Street

Upsal Street takes its name from Upsala, a mansion at 6430 Germantown Avenue, which was probably named for the Swedish city, or its university. Saved from demolition by a citizens' group in the 1940s, the house was built between 1798 and 1801 by John Johnson Jr. (see Johnson Street). The British cannon that bombarded the Chew mansion (see Cliveden Street) was planted on the site of Upsala.

Deeded to the city in 1870 from Germantown Avenue to Chew, Upsal appears in the directory of 1874. From Park Line Drive to Germantown, the street was completed and in use by 1909. In 1923, Sharpnack Street, between Chew and Stenton Avenues, became known as Upsal. Three years later, another stretch of Sharpnack, from the Limekiln Turnpike to Ogontz Avenue, was also changed. The intermediary sections were dedicated to the city between 1924 and 1929.

Valeria Street

See Parrish Street.

Valley Street

See Vine Street.

Venango Street

Taking its name from the Pennsylvania county of the same name (see Counties), Venango Street began its existence as early as 1855, from 2nd Street to 9th. By the turn of the twentieth century, it had grown to extend from Hunting Park Avenue to Mascher Street and from I Street to Balfour.

Venango County took the name of the river, now called French Creek, which flowed through it. The origin of the name is disputed, but it is generally believed to be derived from the Indian *cherango* or *sherango*, which means "bull thistles."

Verbena Avenue

See "Things that Grow . . ."

Verree Road

The Verree family, early settlers in the Bustleton area, operated a grist mill at Pennypack Creek and what is now Verree Road. Robert Verree's mill occupied that site during the Revolution. The area which grew up around his mill became known as Verreeville.

The oldest portion of the road extends from Grant Avenue and Tabor Road to Bloomfield Avenue. That section, dating back to 1820, contains the bed of Old Meeting House Road. Verree Road from Oxford Avenue to Bustleton Avenue—almost the total expanse—was confirmed in 1868. Widened and added to in the twentieth century, the road now extends past Byberry Road.

Vienna Street

See Berks Street.

Vine Street

Vine Street was the original northern boundary of the City of Philadelphia (see Original Streets). The section above the city was known as the Northern Liberties.

The street was originally designated as Valley Street, because it followed the natural depression of a valley. Penn changed the name to Vine, presumably because of the numerous vineyards planted north of the "olde city." Or perhaps the valley had wild vines growing down the sides.

Though it was one of Penn's first designated streets, Vine Street became a legally opened street only in 1883, from the Delaware to the Schuylkill. That portion of the road was widened and converted to an approach to the Schuylkill Expressway in the twentieth century. The section of the street from 52nd Street to Merion Road was confirmed by the Quarter Sessions Court and by affidavit before the turn of the twentieth century.

Vineyard Lane

See Fairmount Avenue.

Vintage Avenue

See Civic Center Boulevard.

Virginia Street

See Wishart Street.

Vogdes Street

Vogdes Street memorializes a prominent Philadelphia family of the nineteenth century. The street first appeared officially under the family's name in 1895, when Gibson's Lane, from Woodland Avenue to Paschall, and Fisher's Avenue, from Market to Vine, were changed to Vogdes. Both Gibson and Fisher were products of the mid-nineteenth century. Vogdes Street, in its entirety, was in use by 1930.

Judging by the timing of the name changes of the two predecessor streets, Vogdes Street probably was named for Jesse T. Vogdes, a prominent West Philadelphia builder, at the instigation, perhaps, of his son Jesse T. Jr. (1858–1919), who had served in the city's survey department for several years before the street was named. In 1898, he was appointed chief engineer and superintendent of Fairmount Park. In this capacity, he supervised the construction of the East River Drive from the Lincoln Monument to the Wissahickon. And, after the Centennial buildings were removed, he took charge of the park's reconstruction, resulting in the beautiful concourse at Memorial Hall.

Whether Vogdes Street was named for the father or the son is immaterial. The contributions of each deserved the honor.

Wakefield Street

Wakefield Street takes its name from Wakefield, the estate of Philadelphia merchant Thomas Fisher. Wakefield Mansion was built about 1795 and named after the Fishers' ancestral home in Yorkshire, England.

The street first appears to have been in use, from East Logan Street to Bringhurst—the site of the house—by 1864. By the end of the nineteenth century, Wakefield Street from East Logan to the dead end past Rittenhouse Street was in public use.

Wakeling Street

See Godfrey Avenue.

Waln Street

According to folk historians, Waln Street was named for Nicholas Waln, one of the first land purchasers in Penn's province. However, because of the street's location and its proximity to Waln Grove, the eighteenth-century country seat of Nicholas's cousin Robert, it seems likely that Robert Waln donated his name to the street.

"Waln Grove was one of the most beautiful seats of that district [Frankford] much favored of old Philadelphia." Robert Waln, in partnership with still another cousin, Jesse, was a prominent Philadelphia merchant. Their dock, between Spruce and Dock Streets, was the first projection of any depth into the river. A member of the Pennsylvania General Assembly, Waln was eulogized as being "more active [than any other contemporary] in his day in all that related to civic and national progress." After his death, his family continued to live at Waln Grove. When the area rapidly became industrialized, the family sold the estate.

The street, as Waln, first appeared in official listings as early as 1864. Once extending from Coral to Frankford, the only section which remains is the earliest part, from Tackawanna to Kinsey.

Walnut Lane

Walnut Lane takes its name from a gigantic black walnut tree which was allowed to remain standing after the road was opened. The earliest record of this lane is in 1767, when it was confirmed by Quarter Sessions Court from Ridge to Henry Avenue. Other sections existed before the twentieth century, including Wayne to Pulaski, Wissahickon Avenue to Lincoln Drive, Germantown to Morton, and Morton to Chew. Walnut Lane in its entirety, from Ridge Avenue to Cheltenham Avenue, was in use by 1936.

Walnut Street

One of Penn's original streets (see Original Streets), Walnut Street was first designated as Pool—because it led to a pool of water at the foot of the street at Dock Creek. Penn quickly changed the name to that of a tree.

The earliest record of lots being laid out along the route of Walnut Street is in 1674/75 on the south side, between 3rd and 4th Streets. By 1716, lots were surveyed from Dock Street to 7th Street, on both sides, and the street was in public use.

Very little information exists on the expansion of Walnut Street from 7th to the Schuylkill River. On the west side of the Schuylkill, the first segment to appear was from 40th Street to the Mill Creek, west of 43rd Street, in 1854. Almost thirty years later (1872), an act of assembly extended the street as far as 57th. This was confirmed by the Quarter Sessions Court in 1878. By 1918, Walnut Street was dedicated to full public use from the Delaware River to Cobbs Creek Parkway.

Walter Street

See Delancey Street.

Wanamaker Street

WANAMAKER STREET. Though the department store that looms in the shadow of Philadelphia's City Hall still carries the Wanamaker name, it is no longer owned by the Wanamaker family. John Wanamaker, founder of the firm, was Philadelphia's "merchant prince" and President Benjamin Harrison's postmaster-general. (Photo circa 1912 by Duhrkoop)

Wanamaker Street honors a native son who became Philadelphia's "merchant prince." In 1861, John Wanamaker (1832–1922) established a small clothing store, which grew to become one of the largest department store operations on the East Coast. After declining the nomination for mayor of Philadelphia in 1886, he was named postmaster-general three years later, under President Benjamin Harrison.

Wanamaker Street first appeared in 1895, when the name of 58½ Street, from Media Street to Lansdowne Avenue, was changed. The next year, probably by coincidence, John Wanamaker opened a larger outlet in New York City. Between that time and 1925, shortly after its namesake's death, Wanamaker Street grew to extend, with several interruptions, from Lindbergh Boulevard to Woodcrest Avenue.

Warder Street

See Blair Street.

Warren Street

Named after Warren County, Pennsylvania (see Counties), Warren Street appeared on street listings as a West Philadelphia street in 1874.

The earliest section (which might have been in public, but not official, use as late as 1857), from Wallace to 41st Street, was vacated in 1960. The second section, Spring Garden to 40th Street, still remains. Another street, Pear—from Master Street to Lansdowne Avenue—was made into Warren at the turn of the twentieth century. All sections of Warren Street were put into public use by affidavit, which seems to indicate that the street grew and was in use much earlier than the records indicate—much the same as the name implies.

Warren County, formed in 1800 out of parts of Allegheny and Lycoming counties, was named for General Joseph Warren (1741–75), who was killed at the

Battle of Breed's Hill, better known, erroneously, as Bunker Hill. The family name Warren means "at or by the warren," the home of rabbits and hares.

Washington Avenue

Commemorating the "father of his country," Washington Avenue was first confirmed by Quarter Sessions Court in 1790, from Swanson Street to Passyunk Avenue. The avenue extended to Broad Street by 1831; to Grays Ferry, by 1839. The western part of Washington Avenue, 57th Street to Cobbs Creek Parkway, was developed from 1905 to 1912. Before 1858, the avenue was called Washington Street from the Delaware River to 3rd Street; from 3rd Street to the Schuylkill River it was known as Prime.

For many years, railroad tracks ran down the center of Washington Avenue—from the Delaware River westward—feeding the industries and businesses which flank the road. Sharp curves, now eliminated, precluded the use of locomotives. So before the tracks were relocated and rearranged, teams of mules were used to pull the cars.

Washington Lane

"Old Germantown ended at Abington road . . . as it led from Roxborough to Abington." As early as 1746, a road—in similar position to today's Washington Lane—is listed as the "Road from Abington to Roxborough." Washington Lane, then called Washington Street, was the northern boundary of old Germantown.

Though there are no official records to indicate when the Abington Road became Washington, it seems likely that the name came into common use after the Battle of Germantown in 1777. Battle maps of the period show that Washington's troops, notably the units commanded by Adam Stephen, Nathanael Greene, and Alexander McDougall, would have come down that road. In fact, not too far off the road, above Germantown Avenue, is where Stephen's men, blinded by the smoke and fog, fired upon Anthony Wayne's detachment.

Washington Lane

Washington Lane, from Wissahickon Avenue to Germantown Avenue, was improved in 1827. The road from Germantown Avenue to Stenton was widened in 1878. There is no indication in the recording of this section of when the road was first put into use.

North Washington Lane, from Limekiln Pike to Cheltenham Avenue, was installed in 1838. The remainder from Stenton Avenue bears no date of confirmation. However, it was widened in 1911 along with the northern section of Cheltenham.

Washington Square

See Squares.

Washington Street

See American Street, Leverington Avenue, Market Street, Washington Avenue, Washington Lane.

Water Street

In 1755, Thomas Pownall wrote that the original idea was that the Front Street on the Delaware would have no houses immediately on the bank, but rather serve as a parapet. "After the first settlers had bought these lots on Front Street," he wrote, "it was found more convenient for the merchants and trades to build their warehouses, and even dwellinghouses, on the beach below, which they wharfed over . . . [and] several took long leases [since the land was not to be sold] and this became a street of the dwelling-houses &c of all the principal Merchants and rich men of business, and was called Water Street."

City records acknowledge that this street existed. In 1883, affidavits were sworn to which indicated Water Street had existed for "at least 21 years."

The original name, given to the street by the early settlers, was "the street under the Bank." Later, when it was laid out in 1705, it was named King Street. From Porter to Christian, the street was known as Otsego until 1895. A small section of Otsego, from Snyder to Morris, was known as Church until 1858. Until their addition brought Water to its full length in 1895, the section from Bainbridge to South was Penn; the section from Cumberland to Gurney was Fox; Clearfield to Lippincott was Emma; Westmoreland to Venango was Fox; and Fisher's Avenue to Clarkson was Wentz.

Wayne Avenue

The first recorded section of this road, from Chesheim Road to Paper Mill, was confirmed in 1746. Possibly, this section is the key to the naming of Wayne Avenue, since Major General "Mad" Anthony Wayne's men, before the Battle of Germantown, were positioned in that general location (see Washington Lane).

The roadway from Tulpehocken to the Philadelphia, Germantown & Norristown Railroad was in use by 1864. By the end of the nineteenth century, Wayne Avenue was in use from Tacoma and Dounton Streets to Carpenter Lane, in addition to the earlier sections. The last dedication took place in 1917—from Ellet Street to Mt. Airy Avenue.

Webster Street

The name of Webster Street is most often assigned to Daniel (1782–1852), statesman and orator, and Noah (1758–1843), lexicographer. But Pelatiah (1726–95) was a political economist and merchant in Philadelphia. He aided in the adoption of the Constitution.

Welsh Road

Laid out in 1711, the road "from the bridge between the land of John Humphreys and Edward Foulke in Gwyneth to the mills on Pennepack creek, as a ford in

Welsh Road

Moreland Township," provided the means for the early settlers to have their grain ground into flour. Because the petitioners were natives of Wales, this road became known as the Welsh Road.

Completed the next year, Welsh Road became the earliest public highway into Horsham Township, separating it from Upper Dublin and Gwynedd. In 1731, after complaints had been received, the Quarter Sessions Court divided the roadway among the three townships. Welsh Road was extended from the Philadelphia—Montgomery County line into the Bustleton area of Philadelphia in 1823.

The remaining sections of Welsh Road, from Cottage Street to the Roosevelt Boulevard, are products of the twentieth century; the last section, from Mower Street to the Boulevard, was dedicated in 1954.

Wensley Street

See Bellevue Street.

Wentz Street

See Water Street.

Westmoreland Street

Named for Westmoreland County, Pennsylvania (see Counties), this street existed—without benefit of dedication—from 17th Street to west of 22nd Street before 1864. Other portions of Westmoreland, from F Street to Balfour, were in regular use by 1887. By 1930, Westmoreland had reached its present distance—35th Street to the Delaware River.

Westmoreland County, formed from Bedford in 1773, was the last county to be organized under the proprietary government. It was named after the English county of Westmorland.

Wetherill Street

See Carlisle Street.

Wharton Street

A "new street" was created by the commissioners of the District of Southwark on January 5, 1790, and called "Wharton street." The road dockets do not indicate for whom the commissioners named it.

Not too far from Wharton Street, Joseph Wharton had his mansion, Walnut Grove, at what is today 5th Street and Washington Avenue. It was in Wharton's home that the famous or infamous "Meschianza" took place during the British occupation of Philadelphia in the winter of 1777. The Meschianza ball was held as a farewell party for Lord William Howe; it drew many local Tories to Walnut Grove. It is highly unlikely that the same men who, in 1790, named a neighboring street "Federal" would have had the broad-mindedness to honor the memory of a man with such strong Loyalist tendencies.

One of the leading Philadelphia merchants of that day, on the other hand, carried the same family name. Thomas Wharton, Jr. (1735–78) served as president of the Supreme Executive Council of Pennsylvania upon its formation under the new state constitution of 1776. During 1775, he was a member of the Provincial Committee of Safety. There was no doubt as to his feelings concerning the Crown. There is some doubt, however, as to the source of the name.

Wheat Street

See Quarry Street.

Wheatsheaf Lane

Taking its name from the small village of Wheatsheaf on

Wheatsheaf Lane

the Bustleton Pike, north of Frankford, Wheatsheaf Lane was in use prior to 1862, from Richmond Street to Frankford Avenue. Widened in 1916, the lane extends as far as it did in the nineteenth century.

Whitby Avenue

Whitby Avenue stands alone. The mansion for which it was named is now miles away. Even the lane that led to it has since disappeared. All that remains is the name.

Whitby Hall, reached by a lane, "heavily shaded by giant sycamores," that led from "the King's Road to the road to West Chester," was built before the Revolution by James Coultas, a native of Whitby, Yorkshire, England. One of the many houses visited by George Washington, Whitby Hall was taken down brick by brick and moved to Haverford in 1922.

The last section of Whitby Avenue was opened in 1890.

WHITBY AVENUE. The mansion for which Whitby Avenue was named was built before the American Revolution by merchant James Coultas. Whitby Hall is nowhere near Whitby Avenue. It was taken down brick by brick in 1922, and moved to Haverford, Pennsylvania.

White Horse Alley

See Bank Street.

Whittell's Lane

See Queen Lane.

Wigard Avenue

Wigard is the English version of an old German masculine name variously spelled *weechert*, *weckhart*, *wiggart*, and *wigart*. The best-known person with that name in the Roxborough-Manayunk area was Wigart Levering (see Levering Street), one of the first landowners in Sommerhausen (now Chestnut Hill).

The earliest road to follow the line of present-day Wigard was a section of Wise's Mill Road, confirmed by the court in 1786, from Shawmont to Henry Avenue. The earliest use of the name was in 1923, in the section from Fowler Street to Ridge Avenue. Wigard Avenue was complete from Hillside Avenue to Henry by 1963, and was widened in 1972.

William Street

See Boudinot Street, Cambria Street.

Willings Alley

Charles Willing, mayor of Philadelphia from 1748 to his death in 1754, lived on this little street, between 3rd and 4th Streets, off Walnut.

The alley dates back to before 1734, when the first Roman Catholic chapel in Philadelphia was established there. The province of Penn was the only area where Catholics felt free to worship in the colonies. Old St. Joseph's Church still occupies that site. The original

WILLINGS ALLEY. Thomas Willing was mayor of Philadelphia from 1748–54. The alley on which he lived, which bears his name, became, in 1734, the home of the first Roman Catholic chapel in Philadelphia. Philadelphia was the only place in the colonies where Catholics were free to practice their faith. Old St. Joseph's Church—in a more modern building—still occupies the site. (Scharf and Westcott's *History of Philadelphia, 1609–1884*)

chapel was replaced in 1757, and the present church was dedicated in 1839.

In 1895, the city sought to change the name of Willings Alley to St. James. Popular historical protest restored the original name two years later.

Willow Grove and Germantown Plank Road

See Mount Airy Avenue.

Willow Street

Pegg's Run, a stream which was at one time navigable as far west as Ridge Avenue and 12th Street, was culverted by the commissioners of Northern Liberties in 1826, from 6th Street to present-day Beach.

Though Willow Street, formerly the bed of Pegg's Run, was opened in 1829, there is doubt that an actual street existed there until after 1870. If Willow Street were laid out and in use as a street, it must have been in a depression in respect to Delaware Avenue. Two ordinances of 1870 appropriated $18,000 to construct a "bridge" at Willow and Delaware Avenue.

The street as such must have been constructed and in public use by 1895. In that year, Centre or Miller Street, from Union to State Street, became Willow.

The original name for Pegg's Run was Cohoquinoque Creek. The stream adopted the name of Daniel Pegg, a brick-maker whose house on Front Street Penn tried, unsuccessfully, to rent on his second visit to his city. The name is perpetuated in Pegg Street. (See Duffield Street.)

Wilmot Street

David Wilmot (1814–68), for whom this street was

named, is best known for offering an amendment to an 1846 House appropriations bill for the settlement of Mexican border claims. The "Wilmot Proviso," as it became known, brought to a head the question of slavery. In it, Wilmot stated that "as an express and fundamental condition to any territory from the Republic of Mexico to the United States . . . neither slavery nor involuntary servitude shall ever exist in any part of said territory." Though the Wilmot Proviso did not pass, it did provide the main basis for the founding of the Republican Party. In June 1862, after the Republicans had gained control of Congress, Wilmot's provisions were approved in an act banning slavery in the territories.

Winfield Street

See Appletree Street.

Wingohocking Street

One of the frequent visitors to James Logan's Stenton mansion was Chief Wingohocking. One day, while standing with Logan by the stream which once wound through the estate, the chief proposed an Indian token of friendship—an exchange of names. Rather than offend the Indian by refusing, Logan responded: "Do thou, chief, take mine, and give thine to this stream which passes through my fields, and when I am passed away and while the earth shall endure it shall flow and bear thy name." Neither Logan, the chief, nor the stream survived. However, the street which was named for the stream gives immortality to Logan's statement, as do Logan Street and Stenton Avenue, which was named for his mansion.

The earliest section of present-day Wingohocking was the old Powder Mill Road, from Nicetown Lane to Adams. That portion of the street was confirmed in 1803. The rest of Wingohocking Street was dedicated from 1890 to 1937.

Wise's Mill Road

This road was named for John Wise, who operated a grist mill on the Wissahicken as early as 1738.

The first recording of the road did not appear on city records under Wise's Mill, but under Wigard Avenue. The road jury viewed the road, from Shawmont to Henry, and Quarter Sessions Court confirmed it in 1786. Apparently the road was extended beyond present-day Henry Avenue in 1812 and widened in 1835. The modern Wise's Mill Road was complete to its present width and distance by 1960.

Wishart Street

L. Q. Cincinnatus Wishart, Philadelphia's last "medicine man" of note, donated his name to this street. Wishart began as a candy dealer in 1847. By 1854, he was a grocer, and six years later he was in the patent medicine business in Kensington, manufacturing a concoction he called Tarcordial.

In 1870, Wishart Court appears in the street lists, followed a decade later by Wishart Street, from Jasper to Emerald. The city did not record this street until 1890. At the time of the uniform naming ordinance of 1895, Wishart Street grew to include Saxton, from Bath to Richmond (now vacated), and Virginia, from Trenton to Amber. In 1897, Park, from 17th to Allegheny, joined L. Q.'s namesake. By 1916, Wishart Street, from Trenton Avenue to 30th Street, was complete.

Wissahickon Avenue

Originally the old Township Line Road which separated Germantown from Roxborough Township, Wissahickon Avenue was built in 1826 along the banks of the creek from which it takes its name, "from the Ridge Road to the Rittenhouse Mill."

The name Wissahickon is a corruption of the Indian name *wisameckham*, meaning "catfish stream." For a brief period, the stream was called Whitpaine's Creek, after an early settler, Richard Whitpaine.

Wissahickon Avenue extended to the Montgomery County line in 1856 and was operated as a toll road by the Wissahickon Turnpike Company. Additional sections of the avenue were incorporated in the early twentieth century; the last segment, in 1955. In 1983, the section of Wissahickon from Forbidden Drive to Ridge Avenue was designated Lincoln Drive.

Wistar Street

See Brandywine Street.

Wistaria Street

During the early nineteenth century, Dr. Caspar Wistar, brother of John Wister (see Wister Street), lived at Grumblethorp in Germantown. Constantly in search of new knowledge, Dr. Wistar entertained many of the notables of his time. One of these visitors, Thomas Nuttall, the British naturalist, named a vine in the doctor's honor. "For euphonious reasons, I shall spell it wisteria," Nuttall told Wistar. Perhaps the naturalist didn't realize his slip in good etiquette in conferring John's family name spelling instead of Caspar's. Regardless, the offspring of the original wisteria (or wistaria) is still blooming behind the former offices of the Mutual Assurance (Green Tree) Company at 4th and Locust Streets.

WISTARIA STREET. Thomas Nuttall, the British naturalist, named a vine in honor of his Philadelphia host: Dr. Caspar Wistar. Unfortunately, Nuttall spelled the name *wisteria*. Caspar's brother, John, spelled the family's name with an *e*; Caspar, with an *a*. (An engraving by Samuel Sartain from a painting by S. B. Waugh after B. Otis)

The City of Philadelphia exercised real tact in the naming of the street. When first deeded in 1926, from Grant Avenue to Murray Street, the name was spelled with an *a*. The remainder of the street, from Fullmer to Lott, Bowler to Hoff, and Cowden to Sandy Road, was deeded to the city between 1946 and 1955.

There are other Philadelphia streets named for fruits, flowers, and plants (see Things That Grow), but none have the historic significance of this street name.

Wister Street

In 1717, Caspar Wüster arrived in Pennsylvania. Four years later, the clerk who recorded his oath of allegiance to the king made a slight mistake and spelled his name Wist*a*r. Caspar apparently liked it because he never attempted to correct it.

When his brother Johann arrived in 1727, however, the clerk was more careful. His name remained Wist*e*r. Johann, or John, was a knowledgeable and competent cultivator of blackberries, and as an end result of his labors, he made and imported wines. By 1744, he had built "Wister's Big House," opposite Indian Queen Lane—the first country seat in Germantown.

Wister Street was laid out and confirmed from the Germantown Road to the Bristol Township line in 1735. Though city records do not list the name for this lane, the 1751 Scull map indicates the line of Wister Street as the "Road leading to the Township Line, commonly called Reayer's Road." Other names attributed to the street have been Duy's and Danenhower's Lanes.

In 1748, the Bristol Township Line Road (now part of Wister) was laid out from Church Street to the Bethlehem Pike. By 1900, Wister Street ran from Germantown Avenue to Stenton. Not until 1927 did the street reach its full length.

Caspar, the incorrectly spelled brother, has not been forgotten—see Wistaria Street.

Wolf Street

WOLF STREET. It's apparent from looking at this illustration that the source of the name for Wolf Street was not a predatory animal. Right? Well, some people think that politicians fit into that category. Regardless, Wolf Street was named for Governor George Wolf.

Contrary to popular local tradition, Wolf Street is not named for an animal that once prowled the lower limits of the city.

George Wolf (1777–1840), the Pennsylvania governor who spearheaded the passage of the Free Public School Act in 1834, is remembered by a street which had its beginnings during the Centennial. Having served as a state legislator and three terms as a member of the U.S. House of Representatives, Wolf was elected governor (see Governors) in 1829. He was defeated for a third term, but politics will out. President Jackson appointed him to the newly created post of comptroller of the treasury in 1836. He resigned that position two years later to become collector of customs for the Port of Philadelphia.

Wolf Street, at its present distance, was dedicated to the city between 1876 and 1928. A small section, from Vare Avenue to 26th Street, was moved to a new location in 1952.

Woodland Avenue

See Baltimore Avenue.

Woodland Street

See Delancey Street.

Wyalusing Avenue

When Penn realized in 1684 that a treaty would be necessary in order to maintain congenial relations with the Indians, he sent emissaries to Chief Tammany (Tamane), the powerful and respected leader of the Delawares. The chief lived in what is now West Philadelphia, and Wyalusing Avenue travels through the chief's ancient home. Some historians indicate this avenue was named for Pennsylvania's Wyalusing Township and/or Wyalusing Creek. At any rate, Wyalusing means "home of the Old Warrior," an apt title for Tammany. Penn, by the way, never met Tammany.

Wyalusing Avenue, first confirmed from Lancaster to Haverford Avenue in 1888, was complete to its present distance by the turn of the twentieth century. In 1895, Sylvan, from 39th to slightly west of the 40th Street dead end, had its name changed to Wyalusing.

Wynne Street

See Chestnut Street.

Wynnefield Avenue

Wynnefield Avenue, from Bryn Mawr to Parkside Avenue, was constructed under order of the Pennsylvania Assembly in 1873 as part of the western approach to the Centennial grounds in Fairmount Park (see Bryn Mawr Avenue). Like Bryn Mawr Avenue, with which it shared the embellishments and decorations of the approach, it was named after the section of Philadelphia—Wynnefield—from which it originated.

Thomas Wynne (1630–92), Penn's "chirurgeon," built his home, Wynnestay, in 1689–90 at the center of five thousand acres along the line of the old Lancaster Road. Anglicized, the name of the estate became Wynnefield.

Following the Centennial, Wynnefield Avenue increased in size—first from Cardinal Avenue to 54th Street, and then from 54th Street to Bryn Mawr Avenue. The entire avenue was complete by 1907.

Wynnewood Road

The name of this street is of fairly modern designation. Gross Street, which was dedicated for use between 1897 and 1910, from Lansdowne Avenue to Woodbine, became known as Wynnewood in 1914. The road follows the line of the suburban Montgomery County thoroughfare of the same name. Unfortunately, the two roads do not connect; there is a several block gap between them. Presumably, the road was named for the suburban community of Wynnewood, which honors Penn's personal physician Thomas Wynne (see Wynnefield Avenue).

Wyoming Avenue

Named for Wyoming County, Pennsylvania (see Counties), this avenue was first deeded to the city, from Rising Sun Avenue to D Street, in 1863—a little more than twenty years after the county was formed in 1842. Before the turn of the twentieth century, the avenue extended from 5th Street to Castor Avenue. The remainder, from 8th to 5th Streets, was completed by 1910.

Wyoming Avenue

Wyoming is a corruption of the Delaware Indian *m'chwewormink*, meaning "extensive plains or meadows." Early settlers, unable to pronounce the word, called the Wyoming Valley "Waioming." It is a short slip of the pen to arrive at the present-day spelling.

York Street

York Street is yet another Philadelphia artery to carry the name of a Pennsylvania county (see Counties). Formed in 1749, York County was named in honor of James Stuart, long-time Duke of York, later King James II. Stuart was a close friend of Admiral Penn, William's father.

The best estimate is that York Street came into being from Richmond Street to 4th Street through the partitions of the Norris and Sepviva estates before 1855. Other sections, from 12th to American Street, were acquired by deed at about the same time. A road jury confirmed a stretch of the street, from old Islington Lane (near 22nd Street) to 24th Street in 1853, but records do not indicate how fast the road was completed. By 1869, the section from 29th Street to near 22nd was dedicated. York Street was completed to its full length—Ridge Avenue to Richmond Street—by the end of the nineteenth century. (See Sansom Street.)

Yower's (Ewer's) Alley

See Black Horse Alley.

Zinkoff Boulevard

Dave Zinkoff, better known to sports fans as "The Voice of the Sixers," died Christmas Day, 1985. By March of 1986, Philadelphia's City Council had renamed Stadium Place in honor of The Zink . . . even though that particular street was not on the city plan.

Zinkoff began his career as a public address announcer when he attended Temple University. He announced every major sport, including the first Sugar Bowl game in 1935. He even covered the Phillies and the Philadelphia Athletics at the same time. He was, however, best remembered for his announcing of the '76ers. Zinkoff was so much a part of that team that the Boston Celtics' Red Auerback said that "visiting teams had to play especially hard in Philadelphia because they were really facing six men—five players and The Zink."

Zoological Street

The land in front of Philadelphia's Zoological Garden was nothing more than swamps following the Centennial of 1876. The improvements to 34th Street, in front of the Zoo, are attributed to the planning of then Chief Engineer Jesse T. Vogdes, Jr. (see Vogdes Street). The other street which circles the Zoo, from 34th to Girard Avenue, was named in 1928 Zoological Street.

Bibliography

Barton, George. *Little Journeys around Old Philadelphia.* Philadelphia: The Peter Reilly Co., 1925.

——. *Walks and Talks about Old Philadelphia.* Philadelphia: The Peter Reilly Co., 1928.

Boatner, Mark M., III. *The Civil War Dictionary.* New York: David McKay Co., 1959.

——. *Encyclopedia of the American Revolution.* New York: David McKay Co., 1966.

Brandt, Francis B., and Henry V. Gunmere. *Byways and Boulevards in and about Historic Philadelphia.* Philadelphia: Privately published by The Corn Exchange National Bank, 1925.

Brey, Jane W. T. *A Quaker Saga.* Philadelphia: Dorrance & Co., 1967.

Bulletin Almanac. Philadelphia: The Bulletin Co., 1924–74. Various volumes.

Burt, Struthers. *Philadelphia: Holy Experiment.* Garden City, N.Y.: Doubleday, Doran & Co., 1945.

Carrington, Henry B. *Battles of the American Revolution.* New York: A. S. Barnes & Co., 1876.

City of Philadelphia. *Ordinances of the Corporation of, and Acts of Assembly Relating to, the City of Philadelphia—1701–1851.* Philadelphia: Crissey & Markley, 1851.

——, Board of Education. *Annual Reports of the Philadelphia Board of Public Education.* Selections made available from the Pedagogical Library of Philadelphia.

——, City Council. *Ordinances and Joint Resolutions of the Select and Common Councils of the Consolidated City of Philadelphia.* Philadelphia: Various printers, 1854–1990.

Bibliography

———, City Planning Commission. *The Consolidated City Plan.*

———, Department of Streets. *Deeds of Transfer, Affidavits and City Solicitor Opinions Relevant to the Legal Status of Street Openings.*

———, Department of Streets. *Legal Status of Streets.* Card file records.

———, Department of Streets. "Preliminary Report: Delaware Avenue Improvement from Dyott Street to Pattison Avenue, History and Legal Status of Delaware Avenue, Including the Legal Status of Intersecting Streets." Undated.

———, Municipal Archives. Road dockets and reports of road juries, Quarter Sessions Court. Allen Weinberg is the Archivist.

Coleman's Reprint of William Penn's Original Proposal and Plan for the Founding and Building of Philadelphia in Pennsylvania, America, in 1683. London: James Coleman, 1881.

Collins, Herman L., and Wilfred Jordan. *Philadelphia: A Story of Progress.* 4 vols. Philadelphia: Lewis Historical Publishing Co., 1941.

The Complete Street Guide to Philadelphia, Pa. Hoboken, N.J.: Geographia Map Co., 1973.

Custis, John Trevor, ed. *The Public Schools of Philadelphia: Historical, Biographical, Statistical.* Philadelphia: Burk & McFetridge, 1897.

Dictionary of Philadelphia. Philadelphia: John Wanamaker, 1887.

Donehoo, Dr. George P. *A History of the Indian Villages and Place Names in Pennsylvania.* Harrisburg: The Telegraph Press, 1928.

Eberlein, Harold, and Horace M. Lippincott. *The Colonial Homes of Philadelphia and Its Neighborhood.* Philadelphia: J. B. Lippincott Co., 1912.

Encyclopedia of American History. (Richard B. Morris, ed.) New York: Harper & Row, 1976.

Espenshade, A. Howry. *Pennsylvania Place Names.* Detroit: Gale Research Co., 1969. Reprint.

Faris, John T. *Old Roads Out of Philadelphia.* Philadelphia: J. B. Lippincott Co., 1917.

Godcharles, Frederic A., ed. *Encyclopedia of Biography.* New York: Lewis Historical Publishing Co., 1931.

Bibliography

Hagner, Charles V. *Early History of the Falls of Schuylkill, Manayunk, etc.* Philadelphia: Claxton, Remsen and Hafflefinger, 1869.

Hamersly, L. R., ed. *Who's Who in Pennsylvania.* New York: L. R. Hamersly & Co., 1904.

Heitman, Francis B. *Historical Register of Officers of the Continental Army.* Baltimore: Genealogical Publishing Co., 1967. Reprint.

Hendrickson, Robert. *Sumter. The First Day of the Civil War.* Chelsea, Michigan: Scarborough House, 1990.

Historical Society of Pennsylvania. *The Pennsylvania Magazine of History and Biography,* 1877–1975. Various numbers.

Hotchkin, Rev. S. F. *Ancient & Modern Germantown, Mount Airy and Chestnut Hill.* Philadelphia: P. W. Ziegler & Co., 1889.

———. *The Bristol Pike.* Philadelphia: George W. Jacobs & Co., 1893.

Jackson, Joseph. *Encyclopedia of Philadelphia.* Harrisburg: The National Historical Association, 1931.

Jones, Horatio Gates. *The Levering Family.* Philadelphia: Privately published, 1897.

Jordan, John W., ed. *Colonial and Revolutionary Families of Pennsylvania.* 3 vols. New York: No pub., 1911.

Keyser, Dr. Naaman H., and others. *History of Old Germantown.* Germantown, Philadelphia: Horace F. McCann, 1907.

Klein, Philip S., and Ari Hoogenboom. *A History of Pennsylvania.* New York: McGraw-Hill Book Co., 1973.

Levering, Col. John. *The Levering Family: History and Genealogy.* Indianapolis: Privately published, 1897.

Lewis, John Frederick. *The Redemption of the Lower Schuylkill.* Philadelphia: The City Parks Association, 1924.

Lynch, Thomas Montgomery, ed. *Encyclopedia of Pennsylvania Biography.* New York: Lewis Historical Publishing Co., 1923.

Maps of Philadelphia and its environs. Located at the Free Library of Philadelphia and the Philadelphia Department of Streets.

Martindale, Joseph C., M.D. *A History of the Townships of Byberry and Moreland.* Philadelphia: George W. Jacobs & Co., n.d.

Bibliography

Morgan, George. *The City of Firsts*. Philadelphia: Historical Publication Society, 1926.

Morris, Charles, ed. *Makers of Philadelphia*. Philadelphia: L. R. Hamersly & Co., 1894.

Nevell, Thomas. "Extracts from John Reed's Book and Measures of the most principal Streets, Squares &c. taken by the Regulators since the year 1782." Unpublished. Probably written about 1784, since the dates listed are much prior to that year; the date in the title is probably a mistake.

Newspaper clippings from the Philadelphia newspapers, especially the *Public Ledger*, the *North American*, the *Bulletin*, the *Philadelphia Inquirer*, and the *Record*. Located in the Philadelphia Free Library, the Library of the Philadelphia Inquirer, the Library of the Bulletin Company, and the Philadelphia Department of Streets.

Pennel, Joseph, and others. *Quaint Corners in Philadelphia*. Philadelphia: John Wanamaker, 1899.

Philadelphia. A 300-Year History. (Russell F. Weigley, ed.) New York: W. W. Norton & Co., 1982.

Rand McNally's Philadelphia Guide. New York: Rand McNally & Co., 1920.

Ritter, Abraham. *Philadelphia and Her Merchants*. Philadelphia: Privately published, 1860.

Rivinus, Marion W. *Lights along the Schuylkill*. Philadelphia: Privately published, 1967.

Rosenthal, Leon S., Esq. *History of Philadelphia's University City*. Philadelphia: West Philadelphia Corp., 1963.

Scharf, J. Thomas, and Thompson Westcott. *History of Philadelphia, 1609–1884*. 3 vols. Philadelphia: L. H. Everts & Co., 1884.

Shackleton, Robert. *The Book of Philadelphia*. Philadelphia: Penn Publishing Co., 1926.

Street directories and telephone books, 1791–1974. Located in the Free Library of Philadelphia and the Historical Society of Pennsylvania.

Vieira, M. Laffitte, comp. *West Philadelphia Illustrated*. Philadelphia: Avil Printing Co., 1903.

Watson, John F. *Annals of Philadelphia, and Pennsylvania in the Olden Times*. 3 vols. Philadelphia: Edwin S. Stuart, 1905. Revisions in Vol. 3 by Willis P. Hazzard.

Bibliography

Webster's American Biographies. (Charles Van Doren, ed.) Springfield, Massachusetts: G. & C. Merriam Company, 1975.

Westcott, Thompson. *The Official Guide Book to Philadelphia*. Philadelphia: Porter & Coates, 1875.

Wildes, Harry Emerson. *The Delaware*. New York: Farrar & Rinehart, 1940.

Acknowledgments
from the 1975 edition, Street Names of Philadelphia

No book of this type would be complete without the singling out of those persons who steered me in the right direction and who assisted beyond the call of duty. Special thanks, then, must be given to the following people. Jerry Post, of the Map Collection of the Philadelphia Free Library, was patient and determined to uncover as much forgotten lore as the Free Library could provide. Allan Weinberg, Archivist of the City of Philadelphia, literally gave me the keys to the Archives so my work could continue after hours. Hazel Anderson and Jesse Mancini of the Philadelphia streets department's Road Records Section contributed some invaluable information. Hazel was intensely interested in the progress of the book and delved deeply into her years with the section to provide anecdotal information. She also provided several helpful volumes from her own personal library. Jesse, once he understood the scope of the project, opened the safe in his office, where I found Nevell's heretofore undiscovered manuscript on early street surveys.

I am also grateful to the entire staff of the Free Library's Government Publications Section. Without the beautiful ladies there, the city ordinances and resolutions would still be on the shelves. As usual, John Platt of the Historical Society of Pennsylvania tolerated me in another of my research escapades. Ben Crane, the dynamic professor of English at Temple University, shared with me many of the papers produced by students in his classes, especially those of Delmas Harris and Laura Taylor, on the subject of streets. Russell F. Weigley read the original manuscript with a historian's

Acknowledgments

eye and an editor's pen. I must also thank the anonymous librarian at *The Philadelphia Inquirer* who let me browse and the many others who, in my research-induced stupor, I ignored, however innocently.

Index

An index is supposed to be a road map, to show the reader what direction the author has taken. But what about a book like this, that lists the contents in alphabetical order? At first we rejected the idea of an index. But, as we began editing and adding, we found that we had created not only a book about the street names of Philadelphia, but also an introduction to cultural communication. An index of the streets, roads, avenues, or other roadways would be superfluous, so we avoided that route. Following are the people, places and things that donated their names to the streets of Philadelphia. When you read about the streets, you will also read about the history of Philadelphia, and the United States. Imagine! All that from a mere index.

A

Abbotsford, 1
Abbot, Charles Frederick, 1
Abington, Pa., 1
Academy of Music, 16
Acadians, 26
Acme Company, 225
"Acres of Diamonds," 62
Adam's Chapel, 112
African Zoar Church, 112
Albany, N. Y., 207
Albright, Jacob, 4
Albright College, 4
Alcott, Amos Bronson, 4
Alcott, Louisa May, 4
Alden, John, 4
Alexander, William, 215
Allegheny County, 5, 63
Allegheny River, 5, 63
Allen, Richard, 2, 194
Allen, William, 6, 161
Allen's Race Course, 120
Allen-Grove Female Seminary, 5
Allentown, Pa., 6
alligew-hanna. *See* Allegheny County.
Altoona Conference of Union Governors, 67
America First Committee, 142
American Museum, 46
American Philosophical Society, 141
American Red Cross, 22
Anderson, Robert, 8
Andre, John, 215
Angora, Village of, 8
Annapolis, Md., 19
Anti-Catholic Riots, 184, 203
Appel, Brian, 175–76
Aramingo, Borough of, 9, 61
Aramingo Canal, 7, 9
Ardmore, Pa., 38
Armat, Thomas Wright, 12
Armat, Thomas, 12
Armat Mills, 11
Armstrong, John, 12
Arnold, Benedict, xx, 17, 97, 146
Arnold, Peggy Shippen, 17, 146
Arthur, Chester Alan, 35, 97
Asbury, Francis, 194
Ashburner, Charles Albert, 13
Ashburton, Baron. *See* Alexander Baring.
Ashburton-Webster Treaty, 21
Ashmead, John, 13
Ashmead, William, 13, 36

Index

Ashmead House, 13
Athensville, Pa. *See* Ardmore, Pa.
Audubon, John James, 14
Auerbach, Red, 249
aufgehende taune. *See* Rising Sun.
Avery the Pirate, 190
Avondale Place, 139

B

Bailed, Francis, 135
Bailey, Lydia R., 16
Bailey, Robert, 16
Bainbridge, William, 17, 186
Baldwin, Matthias William, 17–18
Baldwin Locomotive Works, 17–18
Baltimore, Lord, 176
Baltimore, Md., 116
Baltz, J. & P. Brewing Company, 19
Bancroft, George, *xxii*, 19–20
Baptist Temple, 62
Baring, Alexander, 20–21, 136, 137
Baring Brothers, 21
Barnes, Joseph K., 21
"Barren Fig Tree," The, 112
Barren Hill, 195
Barron, James, 71
Barton, Benjamin Smith, 22
Barton, Clara, 22
Bartram, John, 22
Bartram Gardens, 32
Bartram's Gardens, 80
Basel, Switzerland, 48
Bayard, James Ashton, 23
Beaver, James A., 64, 103
Beaver County, Pa., 64, 103
Beaver, 206
Beck, Paul, Jr., 73
Bedford County, Pa., 64, 236
Belfield Mansion, 24
Bellefield. *See* Belfield Mansion.
Bellefonte Patriot, 174
Bellevue, Pa., 24
Belmont, District of, 25, 61
Belmont Mansion, 25, 149
Belmont Plank Road, 222
Ben-Gurion, David, 25–26
Benezet, Anthony, 26
Berks County, Pa., 63, 224
Berkshire, England, 63
Bethlehem, Pa., 13, 27, 160
Biddle, Clement, 118
Biddle Directory of 1791, 118–19
Bigler, William, 27–28, 103, 183
Bingham, Anne Louisa Willing, 21, 28
Bingham, William, 28, 136, 137

Birch, Thomas, 29
Black Hawk War, 8
Black Horse Tavern (Inn), 29
Blackbeard the Pirate, 190
Blair, John, 30, 64
Blair County, Pa., 64
Blair's Gap, Pa., 30
Blockley Township, 61, 110
Blue Anchor Tavern, 76
Blue Ball Inn, 20
Boar's Head Inn, 20
Bonaffon (farm), 39
Bonaparte, Joseph, 31, 136, 137
Bonsall, Benjamin, 31
Bonsall, Richard, 31
Boone, Daniel, 32, 34
Boonsboro, Ky., 32
Boorse, Jack, 131
Boston Celtics, 249
Boudinot, Elias, 32
Boudinot, Hannah Stockton, 32
Bouvier, John, 33
Bouvier, Michael (Michel), 33
Braddock, Edward, 32, 34
Bradford, Andrew, 34
Bradford, William, 34, 64
Bradford County, Pa., 64
Brandywine, Battle of, 35, 188
Brandywine Creek, 35, 54
Breed's Hill, Battle of, 233
"Brewery-town," 19
Brewster, Benjamin, 35
Bridesburg (Borough), 36, 61
Brill Company, J. G., 36
Bringhurst, George, 206
Bringhurst, John, 36, 206
Bristol, England, 165
Bristol Township, 61
Brumbaugh, Martin G., 103
Bryan, George, 38
Bryn Mawr, Pa., 38, 217, 246
"Bryn Mawr," 38
Bucknell, William, *xxii*, 39, 125
Bucknell University, 39
Bucks County, Pa., 90
Buist, Robert, *xii*, 39
Bunker Hill. *See* Battle of Breed's Hill.
Burbank, Luther, 40
Burnham, England, 40
Burnholme Park, 40
Burnside, Ambrose Everett, 40
Bussil family, *xviii*
Bussil-town, *xviii*
Busti, Paul, 41
Bustleton & Smithfield Turnpike, 222

Bustleton & Sumerton Turnpike. *See* Bustleton & Smithfield Turnpike.
Butler, Richard, 64
Butler County, Pa., 64
Byberry & Andalusia Turnpike, 222
Byberry & Bensalem Turnpike, 111, 222
Byberry Township, 61

C

Cadwalader, John, 42, 78
Cadwalader, Lambert, 42
Cadwalader, Thomas, 42
Cajun, 26
Callahan, David, 8
Camac's Woods, 43
Cambria County, Pa., 63
Cambria Hills, Wales, 44, 63
Camden, S.C., Battle of, 71, 97
Cameron, Simon, 44–45, 64, 174
Cameron County, Pa., 64
Camocks, Sarah Masters, 43
Camocks, Turner, 43
Campbell, John T., 199, 202
Campbell, Rosemary, 199
Campington, 112
Cantrel family. *See* Cantrell.
Cantrell, Francis J., 46
Cantrell, John H., 46
Cantrell, John H., Jr., 46
Carey, Mathew, 46, 155
carkoens. *See* Cobb's Creek.
Carlisle, Cumberland, England, 47
Carlisle, Pa., 47, 75
Carpenter, Ellen, *xvii*, 181
Carpenter, George W., *xvii*, 47, 181
Carpenter, Samuel, 47, 224
Castor, Elwood, 48
Castor, George, 48
Castor, George Albert, 48
Castor, Horace W., 48
Castor, Jacob, 48
Castor, John George, 48
Castor, Thomas, 48
Cathedral of Sts. Peter & Paul, 203
Catherine the Great, 191
Cedar Grove, 165
Cedar Hill Cemetery, 77
Centre County, Pa., 64
ch'ngsessing. *See* Kingsessing.
Chalkley, Thomas, 140

Index

Chambers, Benjamin, 10, 106
Chambers' Ferry, 106
Champlost, Count de, 51
Charleston, S.C., 8
Chaucer, Geoffrey, 52
Chautauga circuit, 62
Cheltenham (Township), 13
Cheltenham & Willow Grove Turnpike, 222
Chelton Hall, 167
cherango. See Venango County.
Chester, Pa., 52–53
Chester County, Pa., 64, 113
Chestnut Hill, 65, 99, 106, 124
Chestnut Hill & Springhouse Turnpike, 27, 99, 222
Chestnut Hill Inn, 106
Chew, Benjamin, 53–54
Chicago, Ill., 24
Chickamaugua, Battle of, 96
"Chinese Wall," 151
Christ Church, 17, 55
Christiana, Queen, *xxv*, 49, 54, 191
Citizens Volunteer Hospital, 218
Civic Center, 225. *See also* Convention Hall
Civic Club of Philadelphia, 20
Clarion County, Pa., 64
Clay, Henry, 66
Claypole, Elizabeth. *See* Betsy Ross.
Clearfield, Pa., 28
Clearfield County, Pa., 63
Clearfield Democrat, 28
Clermont, 95
Cleveland, (Stephen) Grover, 57
Clinton, DeWitt, 57, 64
Clinton County, Pa., 64
Cliveden (mansion), 53, 54, 57, 226
Clooney, Rosemary, 175
Clunie, 146
Clymer, George, 58
coaquannock, xxii
Coates, Thomas, 84
Cobb, William, 58
Cobb's Creek, 58, 150, 177
Cock, Laurens, 208
Cohocksink Creek, 23, 45
Cohoquinoque Creek. *See* Pegg's Run.
College of Physicians of Philadelphia, 177
Columbia County, Pa., 64
Columbia Magazine, 46
Columbus, Christopher, 73–74

Community College of Philadelphia, 59
Convention Hall, 56, 225. *See also* Civic Center
Conway Cabal, 98, 215
Conwell, Russell Herman, 62
Cooconocon Creek, 76
Cooke, Jay, 167
Cooke, Morris Llewellyn, 202
Cornwallis, Charles, 137, 215, 220
Coultas, James, 238
Crawford County, Pa., 64
Credit Mobilier scandal, 97
Crefeld, Germany, 65, 98, 178
Crescent factory, 65
Crescentville, 65
Cresheim. *See* Mount Airy.
Crum Creek, 139
Cumberland, England, 47, 63
Cumberland County, Pa., 63
Curie, Eve, 56
Curie, Marie, 56
Curtin, Andrew Gregg, 66, 103
Custer, George Armstrong, 67
Custer's Last Stand, 67
Czestochowa, Poland, 188
Czolgosz, Leon, 153

Dallas, George Mifflin, 68
Darlington, S.C., 69
Darragh, Charles, 69
Darragh, Lydia, *xxv,* 69
Darwin, Charles, 40
Dauphin County, Pa., 63, 70
Dauphiny Province, France, 70
Davis, Jefferson, 8
Deal, John, 70
Deane, Silas, 71, 188
Decatur, Stephen, 71, 186
DeKalb, Baron. *See* Johann Kalb.
DeLancey, William Heathcote, 72–73
Delaware Bay, 100
Delaware River, 73, 84, 94, 144, 147, 165, 177, 219, 223, 234
Delaware Township, 61
Delaware tribe, 74, 139, 147, 218
Democratic Party, 67, 152
Devereaux, Peter, 74
Dewey, George, 74–75
Diahuga. See Tioga River.
Dick, John, 75

Dickinson, John, 75
Dickinson College, 75
Dictionary of Law, 33
Dixon, F. Eugene, 176
Drew, Daniel, 102
Drexel, Anthony Joseph, 76
Drexel, Francis Martin, 76
Drexel, Joseph William, 76
Drexel & Company, 171
Drexel University, 76, 136
Duffield, Benjamin, 77
Duffield, Edward, 77
Duffield, George, 77
Duffield's Dam, 77
Dunk's Ferry, 77–78
Dunk's Ferry Hotel, 78
Dunlap, John, 78
Duquesne, Fort, 32, 34
Durante, Jimmy, 175
Dyott, Thomas W., 78–79
Dyottville, 78–79
Dyottville Glass-Works, 79

E

Eastwick, 80, 216
Eastwick, Andrew, 80
Edgley, 80, 170
Edinburgh Gardens, 39
Edison, Thomas Alva, 80–81
Elberon, 81
Elfreth, Henry, 81
Elfreth, Sarah Gilbert, 81
Elmwood, 82
Emancipation Proclamation, 67, 142
Engineers Club of Philadelphia, 13
English and High Dutch or German School. *See* Germantown Academy.
Ephrata, Pa., 115
Erie Canal, 57
Erie County, Pa., 64
Erie Railroad, 102
Erwin, Robert, 88
Essex, 17
Essington, 83
Ettwein, John, 56
Evening Bulletin, xxvii

Faire Mount. *See* Fairmount.
Fairmount, 84

261

Index

Fairmount Park, 18, 25, 84, 92, 115, 120, 131, 137, 143, 146, 161, 229, 246
Fairmount Park Commission, *x*, 131, 154, 176, 186
Fairmount Rowing Association, 131
Falls Church, Va., 112
Fare Mount. *See* Fairmount.
Farragut, David Glasgow, 85
Faunce, George W., 86
Faunce, John, 86
Faunce, Thomas, 86
Faunce, William, 86
Fayette County, Pa., 64
Fayette County, Va., 32
Fern Rock, 104
Fields, W. C., 19
Fillmore, Millard, 86
First African United Presbyterian Church, 100
First City Troop, 85, 151
Fisher, Joshua, 89, 100
Fisher, Thomas, 230
Fishtown. *See* Kensington.
Fisk, James, 102
Fitzwalter, Thomas. *See* Thomas Fitzwater.
Fitzwater, Thomas, 90, 142
Flat Rock, 90, 147
Fletcher, Joshua S., 91
Fletcher, Joshua S., Jr., 91
Florence, Italy, 145
Florence, Thomas Birch, 91
Fort Christina, 54
Foulk, Edward, 235
Fox, George, 51
Fox Chase, 86, 92–93
Fox Chase Inn, 92–93
Fraley, Henry, 148
Frankford (Borough), 61, 69, 70, 93–94, 99, 165, 185
Frankford & Bristol Turnpike, 94, 222
Frankford Arsenal, 48
Frankford Creek, 94, 218
Frankford Historical Society, 6
Frankfurt Company, 94
Franklin, Benjamin, *xxii*, 20, 26, 46, 64, 88, 94, 96, 141, 143, 144, 150, 188, 213
Frazier, John W., 4
Frederick Lewis, Prince of Wales, 57
Free African Society, 194
Free Public School Act, 244
French Creek, 227

Frey, Heinrich, 195
Fulton, Robert, 64, 95
Fulton County, Pa., 64, 95

G

Galloway, Joseph, 96
Gandouet, Francis, 212
Garden at Gray's Ferry, The. *See* Gray's Ferry.
Garfield, James Abram, 21, 96–97
Garrett & Eastwick, 80
Gates, Horatio, 97, 215
Geary, John W., 103
General Organization of Jewish Labor, 25–26
Geneva College. *See* Hobart College.
Genoa, Italy, 145
George's Hill, 115
German Township. *See* Germantown.
Germantown (Borough) (Township), 11, 48, 53, 54, 61, 65, 98–99, 114–15, 129, 133, 142, 154, 156, 161, 165, 178, 181, 187, 206, 224, 242
Germantown, Battle of, 111, 188, 233, 235
Germantown Academy, 36, 143, 144, 206
Germantown & Perkiomen Turnpike, 111, 195, 222
Germantown & Reading Turnpike, 99, 222
Germantown & Willow Grove Plank Road, 222
Germantown Union School House. *See* Germantown Academy.
"Germantown Wagon," 36
Gerster family. *See* Castor.
Gilbert, John, 81
Gillingham, Joseph J., 99
Girard, Stephen, 52, 59, 73, 99–100
Girard Bank. *See* Mellon Bank.
Girard College, 49, 58–59, 99, 100
Glen Fern, 6
Glengarry, 223
Gloucester, John, 100
Godfrey, Thomas, 100
Gold's Hotel, 106
Goode, W. Wilson, 74
Goosetown, 197
Gorgas, Joseph, 101, 158–59
goschgosch. *See* Hog Island.

Goshen, Pa., 113
Gould, Jason (Jay), 102
"Governor's Woods," The, 197
Gowen, Franklin B., 104
Gowen, James G., 104
Grace Baptist Church, 62
Grange Farm, 104
Grant, Samuel, 105
Grant, Ulysses S., *xxviii*, 67, 85, 104–5
Gratz, Edward, 105
Gratz, Michael, 105
Gratz, Rebecca, 105
Gratz, Simon, 105
Graver, John, 106
Gray, George, 106, 108
Gray's Ferry, 106–7
Greeley, Horace, 67
Greenberg, Fredavid, 180
Greene, Nathanael, 233
Greene County, Pa., 64
Gregg, Andrew, 103
Grubb, Curtis, 108
Grubbtown, 65
Grumblethorp (mansion), 243
Guiteau, Charles J., 97
Gunner's Run, 9, 23
Gunner's Run Improvement Company, 9
Gwynned, Pa., 235, 236

H

Haddington, 110
Hagy (Hagge), Jacob, 110
Haines, Caspar, 111
Haines, Margaret Wister, 111
Haines, Reuben, 111
Haldeman, Ben, 111
Hamilton, Andrew, 33
Hamilton, James, 53
Hamilton, William, 33
Hamilton Village, 53, 150
Hancock, John, 137
Harmony Fire Company, 112
Harris, Thomas, 21
Harrison, Benjamin, 232
Harrison, Thomas Alexander, 112
Harrison, William Henry, 66, 112
Harrison, Winans & Eastwick, 80
Harrowgate Springs, 132
Hartranft, John F., 103
Hastings, Daniel H., 103
Haverford, Pa., 238
Haverford Meeting House, 113
Haverford Township, Delaware County, 113

Index

Haverford-West, Pembrokeshire, England, 113
Hayes, Rutherford B., 153
Hazzard, Ebenezer, 114
Hazzard, Erskine, 114
Hazzard, Samuel, 114
Hermits of the Mystic Brotherhood. *See* Hermits of the Ridge.
Hermits of the Ridge, 114–15
Heston family, 116
Hestonville, 115–16
Hibernian Society, 155
Hickorytown, Pa., 99
Hicks, Elias, 116
Hicks, Nathan, 93
Hicksites, 116, 171
Hiester, Joseph, 103
Historical Society of Pennsylvania, 20
History of the United States, 20
Hoagie, 117
Hobart College, 73
Hog Island, 117
Hogies. *See* Hoagie.
Holg, Israel, 117
Holme, Elenor, 118
Holme, Thomas, *xxiv*, 2, 10, 47, 84, 118, 127, 173, 192
Holmesburg, *xxiv*, 117–18
Homer, Irv, 175
Hornet, 17
Horsham Township, 236
Hosier, Harry ("Black Harry"), 112–13
Houston, Sam, 185
Howe, William, 69, 96, 237
Hoyt, Henry M., 103
Hudson, Henry, 73
Hudson River, 95
Huguenots, 26
Humphreys, John, 235
Humphreysville, Pa. *See* Bryn Mawr, Pa.
Hunting Park, 120
Hunting Park Race Course, 120
Huntingdon, Countess of, 63
Huntingdon County, Pa., 63, 184

I

Independence Mall, 122
Independence National Historical Park, 141
Independence Square, 213
Indian Queen Hotel (Tavern), 123, 190

Indian Run Creek, 110
Indiana County, Pa., 124–25
Indiana Land Company, 125
Ingersoll, Jared, 125
Innocents Abroad, 127
Institutes of American Law, 33
Insurance Company of North America, 114
Iseminger, Adam, 125
Iseminger, Schubert, 126
Ivanhoe, 105

J

Jackson, Abraham Reeves, 127
Jackson, Andrew, 127, 244
James II, King. *See* James Stuart.
Jansen, Dirck, 129
Jefferson, Thomas, *xxii*, 64, 78, 127
Jefferson County, Pa., 64, 127
Johnson, Andrew, 128
Johnson, John, 129
Johnson, John, Jr., 226
Johnson, Rachel Livezey, 129, 144
Johnson, Samuel, 129
Johnston, William F., 103
Jones, Absalom, 2
Joseph, 195
Juniata County, Pa., 64, 129–30
Juniata. *See* Juniata County.

K

Kagge, William, 110
Kalb, Johann, 71
Kansas Pacific Railroad, 102
karakung. *See* Cobb's Creek.
kararikung. *See* Cobb's Creek.
Katz, Henry, 110
Keenan, Mike, 179
Kelly, John B., *x*, 131
Kelly, John B., Jr., *x*, 131, 176
Kelly, Joseph, *xii*, 180
Kelly, Princess Grace of Monaco, *x*, 131
Kelpius, Johannes, 114–15
Kennedy, Jacqueline Bouvier, 33
Kennedy, John F., *xi*, 128
Kennedy Plaza, 128
Kensington (District), 30, 61, 79, 94, 113, 132, 208
Kensington & Frankford Plank Road, 222
Kensington & Oxford Turnpike, 196, 222
Keyser, Dirck, 117, 132–33

Kidd, Captain, 190
Kingsessing, 31, 75, 133, 191
Kirkbride, Joseph, 36
Knight, Giles, 134
Know-Nothing Party. *See* Native American Party.
Knox, Henry, 134
Kochesberger Family, 108
Kriegsheim. *See* Kriesheim.
Kriesheim, Germany, 65

L

Lafayette, Marquis de, 46, 48, 64, 71, 111, 205
Laine, Frankie, 175
Lancaster, Pa., 101, 136
Lancaster County, Pa., 64, 135, 148, 224
Land, Ann, 210
Lansdowne, 21, 136–37
Lansdowne, Marquis of. *See* William Petty.
Lansdowne House, London, 136, 137
Lanza, Mario, *xi*
Latrobe, Benjamin Henry, 204
Laurel Hill, 96
Laurel Hill Cemetery, 92
Laurens, Henry, 137
Laurens, John, 137
Lawrence County, Pa., 64
Leamy Estate, 132
LeBourget Airfield, Paris, 142
Lebanon County, Pa., 64
lecha. *See* Lehigh River.
lechanwekink. *See* Lehigh River.
Lee, Charles, 138, 215
Lee, Francis Lightfoot, 138
Lee, Henry "Light Horse Harry," 138
Lee, Richard Henry, 138
Lee, Robert Edward, 138
Leech, Tobias (Toby), 13, 93
Lehigh County, Pa., 63, 138–39
Lehigh Navigation Company, 114
Lehigh River, 63, 139
Leiper, Thomas, 139
Lenni-Lenape tribe, 117, 119, 162, 195
Lesley, Joseph, 38
Letitia Street House, 140
Levering, Gerhard, 140
Levering, Wigard (Wigart), 140, 239
Levering, William, 140
Levering School (William), 140

263

Index

Leverington Hotel, 140
Liberty County, Pa. *See* Fulton County, Pa.
Library Company of Philadelphia, 141
Life of George Washington, 46
Limekiln Turnpike, 142, 222, 226
Lincoln, Abraham, 8, 21, 44, 66, 142
Lindbergh, Charles Augustus, 142
Lindbergh, Pelle, *xii*, 179–80
Linnaeus, Carolus, 22
Little Big Horn River, 67
Little Tacony Creek, 77, 99, 218
Little Women, 4
Livezey, Thomas, 6, 143, 144
Logan, James, 100, 144, 213, 214, 241
Lombards, 145
Loudon County, Va., 12
Loudon Mansion, 12
Louis, the Dauphin, 62, 70
Louis XVI, 63, 70
Lower Dublin Academy, 2
Lower Dublin Public School, 2
Lower Dublin Township, 61, 118, 223
Lower Ferry. *See* Chambers' Ferry.
Lukens family, 185
Lutheran Church, 160
Luzerne County, Pa., 64
lycomin. See Lycoming County.
Lycoming County, 64, 145, 185
Lycoming Creek, 64, 145
Lycoming Gazette, 174

M

Macalester, Charles A., 105, 146, 223
Macpherson, John, 118, 146
Madery, Jacob, 156
Maghee, William, 15
Maine, 153
Manatawny Creek, 194
Manayunk (Borough), 61, 90, 147
Manchester, Duke of, 28
Manheim, Germany, 148
Manheim, Pa., 148
Manilla Bay, 74
Mantua, Italy, 149
Mantua Village, 41, 149
Markoe, Abraham, 85, 151
Mascher, John F., 119
m'cheueuwormink. See Wyoming County.

McClellan, George Brinton, *xxii*, 152
McClure, Alexander K., 44
McDougall, Alexander, 233
McKean, Thomas, 64, 75, 103, 153
McKean County, Pa., 64, 153
McKinley, William, 153
McMahon, David, 153–54
McMichael, Morton, 154
McNulty, Thomas J. "Reds," 154
Meade, Fort George Gordon, 155
Meade, George, 155
Meade, George Gordon, *xxii*, 40, 155
Meade, James, 52
Mechanicsville, 155
Medary, Sebastian, 156
Medary, William F., 155
Medical College of Pennsylvania, 1
Mellon Bank, 99
Melrose Abbey, 1
menateyonk. See Manayunk.
mene-iunk. See Manayunk.
Mennonite, 65, 132–33, 197, 198
Mercer, Fort, 93
Mercer County, Pa., 64
Mermaid Hotel, 156
Mershon, Abner H., 156
Mershon, Charles O., 156
Mershon, Cornelius, 156
Mershon, George, 156
Mershon Patent Shaking Grate Works, 156
Meschianza, The, 237
Mifflin, Fort, 93, 157, 162
Mifflin, Thomas, 64, 103, 157
Mill Creek. *See* Cobb's Creek.
Miller Estate, 104
Milner, Edward, 101, 159
Minneapolis, Minn., 62
Minneapolis Daily Chronicle, 62
Minuit, Peter, 54
Missouri Pacific Railroad, 102
Mobile Bay, Ala., 85
Molly Maguires, 104
Monastery, The, 101, 158, 159
Monastery of the Wissahickon. *See* The Monastery.
Monmouth Court House, Battle of, 215
Monodhone Creek. *See* Paper Mill Run.
Monroe County, Pa., 64, 195
Montgomery County, Pa., 13, 64, 141–42, 236, 246
Monument Cemetery, 205

Moore, Cecil B., *xi*, 49–50
Moravian Church, 160
Moreland Township, 61, 220, 236
Morris, Robert, 28, 204
Morris, Stephen P., 218
Morris' Hill. *See* Fairmount.
Morse, Samuel F. B., 183
Moscow, Russia, 8
Mother Bethel A.M.E. Zion Church, 194
Mount Airy, 65, 161
Mount Moriah Cemetery, 50
Mount Pleasant, 146
Mount Vernon, Va., 107
Moyamensing (District), 162
Moyamensing Prison, 162
moyamensing. See Moyamensing.
Mud Island. *See* Fort Mifflin.
Mullen, Priscilla, 4
Municipal Services Building, 128
Munster, Germany, 140
Mutiny of the Pennsylvania Line, 193
Mutual Assurance Company ("The Green Tree"), 243

N

Native American Party, 28, 183
Nazareth, Pa., 160–61
Neumann, John Nepomucene, 203
Nevell, Thomas, 168–70
New York, N.Y., 107
Newton, Isaac, 100
Nice (Neus) family, 165
Nicetown, 165
Noble family, 165
Normandy Village, *xxiii*
Norris, Isaac, 166, 204, 224
Norristown, Pa., 195
North, Oliver, 71
North American, *xxvii*, 154
Northampton County, Pa., 64, 139
Northern Liberties (District) (Township), 61, 228
Northumberland County, Pa., 64, 145
Noyes, Alexander Dana, 102
Nuttal, Thomas, 243

O

Oak Lane, 167
Observations on the Slavery of the Africans and Their

Index

Descendents, 116
Ogden, George, 108
Ogontz, Chief, 167
Ogontz (mansion), 167
"Old Ironsides," 17
Old Pine Street Presbyterian Church, 16, 77
Old St. Joseph's Church, 239–40
Olympia, 75
Orion, William, 170
Orion Tract, 170
Ormandy, Eugene, 16
Ormiston (mansion), 96
Orthodox Friends, 171
Overbrook, 171
Overbrook Farms, 171
Oxford Plank Road, 222
Oxford Township, 48, 61

P

pachsegink. See Passyunk.
pachsegonk. See Passyunk.
Packer, William F., 103, 174
paisajung. See Passyunk.
Palestine, 25, 26
Palmer, Anthony, 132
Palmer Burying Ground, 113
Palumbo, Antonio, 175
Palumbo, Frank, *x*, 175–76
Palumbo, Kippee, 176
Palumbo's Restaurant & Night Club, 175
Park Towne Place, 43
Parrish, Joseph, 176–77
Paschal Iron Works, 218, 219
Paschall, Thomas, 177
Paschallville, 177
passajon. See Passyunk.
passajungh. See Passyunk.
passayunk. See Passyunk.
passuming. See Passyunk.
passyonk. See Passyunk.
Passyunk (Township), 61, 178, 191
Pastorius, Francis Daniel, 98, 178
Pattison, Henry M., 103
Paxtung Township, Lancaster County. See Dauphin County.
Peale, Charles Willson, 24
Peart, Thomas, 10
Peel Hall, 59
Pegg, Daniel, 240
Pegg's Run, 240
Penn, (Admiral) William, 248
Penn (District) (Township), 61
Penn, Hannah Callowhill, *xxiv*, 42, 43

Penn, John, 136, 137
Penn, Letitia, 140
Penn, Richard, 43
Penn, William, *xiii, xxiii, xxiv, xxv, xxix,* 11, 37, 42, 43, 47, 53, 54, 76, 90, 94, 111, 118, 124, 134, 135, 140, 150, 166, 167, 181, 186, 192, 197, 208, 209, 211, 212, 213, 216, 219, 228, 240, 245
Penn Treaty Park, 208
Pennsylvania Agricultural Society, 186
Pennsylvania Herald, 46
Pennsylvania Hospital, 41, 57, 177
Pennsylvania Intelligencer, 174
Pennsylvania Packet, or The General Advertiser, The, 78
Pennsylvania Railroad, 24, 38, 43, 51, 105, 108, 128, 151, 183, 225
Pennypack Creek, 94, 235
Pennypacker, Samuel W., 103
Perkiomen Creek, 195
Perkiomen Turnpike, 99
perlajungh. See Passyunk.
Peters, Richard, 24, 149
Peters, William, 24
Petersburg, Va., 40
Petty, William, 28, 136, 137
Petty's Island, 147
Philadelphia, 17, 71
Philadelphia Almshouse, 57, 177
Philadelphia Athletics, 249
Philadelphia, Bala & Bryn Mawr Turnpike, 222
Philadelphia, Brandywine & New London Turnpike Company, 18, 19, 222
Philadelphia, Bristol & Morrisville Turnpike, 222
Philadelphia, County of, 60
Philadelphia Flyers, *xii*, 179
Philadelphia, Germantown & Norristown Railroad, 18, 235
Philadelphia Inquirer, The, xxvii
Philadelphia & Lancaster Turnpike, 18, 223
Philadelphia & Lancaster Turnpike Corporation, 28
Philadelphia Light Horse. See First City Troop.
Philadelphia Medical Society, 177
Philadelphia Museum of Art, 84, 131
Philadelphia Naval Base, 138
Philadelphia Naval Shipyard, 102, 138

Philadelphia Phillies, 249
Philadelphia Society for Promoting Agriculture, 25
Philadelphia Society for the Promotion of National Industry, 46
Philadelphia, Wilmington & Baltimore Railroad, 108
Philadelphia Zoological Society (Zoo), *x*, 175, 249
Philadelphus, Ptolemy, *xxiv*
Phil-Ellena (mansion), *xvii*, 181
Physick, Philip Syng, 80, 193
Pike, Zebulon, 64
Pike County, Pa., 64
Pittsburgh, Pa., 24, 34
Plattenbach, Joseph, 195
Plonsk, Russian-Poland, 25
Plumbers Union, Local 690, 154
Plymouth Meeting, 194
Poinsett, Joel, *xii*
Polk, James K., 68
Pollock, James, 66, 103, 182–83
poqueski. See Poquessing Creek.
Poquessing Creek, 94, 105, 223
Port Royal House, 185
Porter, David R., 103, 184
Potter, James, 185
Potter County, Pa., 64, 185
Pottstown, Pa., 99
Powel, John Hare, 186
Powel, Samuel, 186
Powell, John, 113
Powell's Ferry, 113
Powelton estate, 186
Pownall, Thomas, 94–95, 98, 234
Preble, Edward, 186
Price, Eli Kirk, 61–62, 146, 186–87
Priestley, Joseph, 188
Public Inquirer (Sunbury, Pa.), 174
Public Ledger, xxvii, 57
Pulaski, Casimir (Kazimierz), *xxii,* 188–89
Pulitzer Prize, 142
Purdon, John, 98

Q

Quary, Robert, 190
Queen Anne, 191
quistconk. See Hog Island.

R

Rabbath-Ammon, *xxiv*
Randolph, Jacob, 192–93

Index

Randolph, Sarah Emlen Physick, 192
Ranier, Prince, 131
Rawle, Francis, 10
Reading, Pa., 99
Record, xxvii
Redfield, Anna Maria, 4
Reed, Joseph, 193
Republican Party, 67, 241
Richmond (District), 61, 209
Ridge Avenue Turnpike Company, 195
Ridge Pike, 223
Ridley Creek, 139
Rising Sun (village), 195–96
Rising Sun Inn (Tavern), 196
Ritner, Joseph, 103, 196
Rittenhouse, David, 32, 197, 198, 213
Rittenhouse (Rittenhousen), William, 197–98
Robin Hood Dell, 92
Robin Hood Ford, 92
Robin Hood Tavern, 92
Robinson, Patrick, 107
Rocksborrow. *See* Roxborough.
Roosevelt, Eleanor, *ix*
Roosevelt, Franklin Delano, *ix*
Roosevelt, Theodore, *ix*, 153, 199
Roosevelt Field, Long Island, New York, 142
Roset, Jacques, 148
Ross, Betsy, 50
Roxborough, 110, 140, 141, 199, 242
Roxborough School, 140
Rubican, Allen, 200
Ruffner, Elizabeth, 200
Ruffner, William Anthony, 200
Ruffner, William Anthony, Jr., 200
Rush, Benjamin, 32
Ryerss, Robert Waln, 40

'76ers, 249
7th Cavalry, 67
sachamexin. See Penn Treaty Park.
St. Clair, Arthur, 207
St. George's Methodist Church, 194
St. Peter's Episcopal Church, 71, 72
St. Petersburg, Russia, 80
St. Stephen's Roman Catholic Church, 200

St. Thomas Protestant Episcopal Church, 2
Saloman, Haym, 25
Sandusky, Ohio, 167
Sansom, William, *xxvi*, 89, 204
Saratoga, Battle of, 97
Sartain, John, 204–5
Sartain, Samuel, 243
Savannah, Siege of, 189
Schulze, John A., 103
Schuyler, Philip John, 207
Schuylkill County, Pa., 64
Schuylkill Navigation Company, 147
Schuylkill River, 64, 84, 131, 136, 144, 147, 150, 177, 194, 207
Scott, General Winfield, 21, 66
Scott, Sir Walter, 1, 105
Scull, Nicholas, 47
Seventh-Day Baptists, 101, 158, 159
Seward, William H., 21
Shackamaxon, *xxiii, xxiv*, 208
Shasta daisy, 40
Shawnee tribe, 32
Sheetz, Henry, 110
Shelburne, Earl of. *See* William Petty.
sherango. See Venango County.
Sherwood Forest, 8
Shiloh, Battle of, 96
Shippen, Edward, *xx*, 17, 148
Shippensburg, Pa. 17
Shisler, Augusta Pulch, 3
Shisler, George W., 3
Shoemaker, Thomas, 143
Shunk, Francis R., 103
Shute (Swanson), Swen, 48, 191
Sign of the Tun Inn, 140
Sisters of Notre Dame, 203
Sisters of the Third Order of St. Francis, 203
Sligo Iron Works, 184
Snyder, David, 8
Snyder, Simon, 64, 103
Society of Friends, 113, 116, 171
Somerdale Borough, N. J., 180
Somerhausen. *See* Chestnut Hill.
Somerset County, Pa., 63
Somersetshire, England, 63
Somerton, 211
Somerville Journal, 62
Songhurst, John, 192
Southwark (District), 61, 95, 190, 191, 193, 211
Spanish-American War, 74, 153
Spirit of St. Louis, The (book), 142

Spirit of St. Louis, The, 142
Spring Garden (District), 61, 211–12
Spruce Mills, 220
Standish, Miles, 4
Stanton, Edwin M., 21
Stenton (mansion), 100, 214, 241
Stephen, Adam, 233
Stiles, Edward, 185
Stiles family, 185
Stille, Olaf, 162
Stirling, Lord. *See* William Alexander.
Stockton, Richard, 32
Stokley, Horatio N., 215
Stokley, William S., 215
Stone, William A., 103
Strettell, Philatesia, 165
Stroup, Henry, 220
Stuart, Edwin S., 103
Stuart, Gilbert, 28
Stuart, James, 63, 248
Sumter, Fort, 8
Sunbury & Erie Railroad, 174
Sunville, 195
Susquehanna County, Pa., 63
Susquehanna River, 63
Swanson, Catharine, 49

Tacony Creek, 218
Tacony Farm, 48
Tacony Plank Road, 223
Tacony-Palmyra Bridge, 218
Tamane, Chief. *See* Chief Tammany.
Tammany, Chief, 195–96, 245
Tarcordial, 242
Tasker, Thomas J., 218–19
Tate, James H. J., 128
tavego. See Tioga River.
Taylor, Zachary, 66
Tecony (Tekone) Creek. *See* Tacony Creek.
Temple College. *See* Temple University.
Temple University, 62
Tener, John K., 103
Thomas, Daniel, 200
Ticonderoga, Fort, 207
Tilghman, Richard Albert, 220
Tilghman, Tench, 220
Tinicum Island, 83
Tioga County, Pa., 64, 221
Tioga River, 64, 221
Torresdale, 223

Index

Torrisdale, Scotland, 223
Torrisdale (estate), 223
Trappe, Pa., 99
Trent, William, 224
Trenton, N.J., 78, 217, 224
Tripoli, 71
Tulpehocken Township, Berks County, 224
tumanaraming. See Gunner's Run.
Twain, Mark, 127
Twining, Thomas, 107–8

U

Union County, Pa., 64
Union School House of Germantown. See Germantown Academy.
Unitarian Church, 188
U. S. Naval Academy, 19
U. S. Naval Asylum, 95
USS *Constitution* ("Old Ironsides"), 17
University of Lewisburg. See Bucknell University.
University of Pennsylvania, 21, 72, 105, 225
Upland, England, 170
Upland, Pa. See Chester, Pa.
Upper Dublin, 142, 236
Upsala, Sweden, 226
Upsala, University of, 226
Upsala (mansion), 226

V

Valley Forge, Pa., 93
Vanderbilt, Cornelius, 102
Variations of Animals and Plants Under Domestication, 40
Venango County, Pa., 64, 237
Venango River, 64, 227
Verna, Anna C., 73
Verree, Robert, 227
Verreeville, 227
Virgil, 149
Vodges, Jesse T., 229
Vodges, Jesse T., Jr., 229, 249
Von Stiegel, Baron Heinrich Wilhelm, 148

W

waioming. See Wyoming County.

Wakefield (mansion), 230
Walking Purchase of 1737, 124
Waln, Jesse, 230
Waln, Nicholas, 230
Waln, Robert, 77, 230
Waln Grove, 77, 230
Walnut Grove (mansion), 237
Wanamaker, John, 232
Wanamaker's, 232
Warren, Joseph, 232–33
Warren County, Pa., 64, 232
Washington, George, *xxii, xxv,* 28, 34, 36, 64, 69, 71, 77–78, 93, 97, 107, 157, 188, 193, 205, 207, 213, 215, 220, 233, 238
Wayne, Anthony, *xxii*, 64, 188, 233, 235
Webster, Daniel, 21, 235
Webster, Noah, 235
Webster, Pelatiah, 235
Wechhart. See Wigart Levering.
Weechert. See Wigart Levering.
Weed, George, 108
Weems, Mason Locke, 46
Welcome, 90, 134
West, Benjamin, 209
West, William, 73
West Chester, Pa., 238
West Philadelphia (District), 53, 61
West Philadelphia Homestead Association, 91
West Point (U.S. Military Academy), 67, 152
Western Avenue, Chicago, 38
Western New York, Diocese of, 73
Westmoreland, England, 63, 236
Westmoreland County, Pa., 63, 236
Wharton, Joseph, 237
Wharton, Robert, 144
Wharton, Thomas, Jr., 237
Wheatsheaf (village), 237
Whitby, Yorkshire, England, 238
Whitby Hall, 238
White, William, 72
White Hall (Borough), 61
Whitemarsh, *xxv*, 110
Whitpaine, Richard, 243
Whitpaine's Creek. See Wissahickon Creek.
Wiccaco, 191
Wigart. See Wigart Levering.
Wiggart. See Wigart Levering.
Willing, Charles, 239

Willow Grove & Germantown Plank Road, 161
Wills Eye Hospital, 177
Wilmot, David, 240–41
Wilmot Proviso, 241
Wilson, S. Davis, 225
Wingohocking, Chief, 144, 214, 241
Wingohocking Creek, 12, 241
wisamechham. See Wissahickon Creek.
Wise, John, 242
Wise's Mill, 242
Wishart, L. Q. Cincinnatus, 242
Wissahickon Creek, 101, 105, 114, 158, 194, 195, 242–43
Wissinoming Park, 77
Wistar (Wuster), Caspar, 177, 243, 244
Wister, John (Johann), 243, 244
Wister, Sarah Logan, 24
Wister, William, 24
Wolf, George, 103, 244–45
Women of the Wildernes, 115
Woods, Georgie, 49–50
Woodvale. See Camac's Woods.
Woodward, George, 202
Wyalusing County, Pa., 245
Wyalusing Creek, 245
Wyck (Mansion), 111
Wynne, Thomas, 53, 246
Wynnefield, 246
Wynnestay, 246
Wynnewood, 246
Wyoming County, a., 64, 246–47

Y

York, Duke of. See James Stuart.
York County, Pa., 63, 248
Yorkshire, England, 230
Yorktown, Va., 215, 220

Z

Zinkoff, Dave, 249
Zinzindorf, Count Nicholas, 13, 160
Zoological Gardens. See Philadelphia Zoological Society.

267